Denver Art Museum
August 18-September 30, 1979

Selections from the Lutz Bamboo Collection

by Ronald Y. Otsuka

Walter E. Lutz

Library of Congress Catalog Card Number: 79-54232
ISBN 0-914738-16-X
Printed in the United States of America

Design: Amelia Ives
Photographs: Lloyd Rule, with the exception of
no. 154 by Paul V. Newman
Printing: A. B. Hirschfeld Press

Cover detail and title page: Brushpot with the
"Seven Sages of the Bamboo Grove" (43)
Inside cover detail: Asai Tonan, *Bamboo* (228)

Foreword

4

Bamboo is a simple plant that belies the diversity of uses and appearances to which it lends itself. Nowhere else in private hands can one find a more comprehensive and brilliant revelation of the possibilities of bamboo as a medium for the artist than in the collection of Walter and Mona Lutz. With unflagging interest over a 35-year period, they have assembled more than 2,000 examples of bamboo art into what has become both the largest and the finest privately owned collection of bamboo objects in the world. The Denver Art Museum is deeply honored and proud to house the collection and to present the first in a series of five annual exhibitions planned to demonstrate the richness of the holdings.

This exhibition attempts to encompass the remarkable scope of the collection and to emphasize its consistently high quality. The 230 objects on display, approximately ten percent of the total collection, range from imaginative and delicate utilitarian tools and baskets to exquisitely carved brushpots and other accouterments of the scholar's studio. Reflecting the highest ideals of the literati, the superb Chinese carvings are one of the principal strengths of the collection and often reveal stylistic parallels with Chinese carvings in other media, such as jade, ivory, and boxwood. Another significant strength is the large group of objects from Japan used for the tea ceremony and flower arrangements. Spanning four centuries and consisting mainly of objects from China, Korea, Japan, and Thailand, the collection also includes a few rare pieces from Africa, India, and the Philippines.

No less remarkable than their collection are the Lutzes themselves. Walter's initial enthusiasm for collecting bamboo objects has been avidly shared by Mona, a master of the Ohara school of Ikebana and a past president of Ikebana International, their two daughters, and their son-in-law. They are to be congratulated for their great success in forming this important collection. We are profoundly grateful to them for making possible the present exhibition and catalog. Future exhibitions will focus on different facets of the collection and will afford viewers a unique opportunity to study the many fascinating aspects of bamboo art.

Thomas N. Maytham
Director

15

Collector's Preface

The question I am most often asked is, How did you begin to collect bamboo objects? Implied in that is another question, Why did you collect them? for it is admittedly a strange occupation, especially for an American whose early acquaintance with this giant grass was limited to walking canes, poles for vaulting and fishing, and porch shades.

Simply put, had there been no World War II, there would have been no collecting of bamboo objects. It all began in the Philippines in 1945 when, as a lieutenant in the Army Corps of Engineers, I had to scout around for materials to supplement our GI stocks for various building purposes. The easiest material to find was bamboo, abundant throughout the Philippines and ours for the cutting on our base island of Leyte.

My affection for the plant at that time was tempered by the many cuts I sustained when I tried splitting poles in half without proper caution or technique. I also soon discovered that the best-designed projects came to naught if one disregarded proper seasoning methods. Powder-post beetles would attack and leave a riddled mess in hours. One such project was bamboo plumbing for the base latrine: the concept was solid, the end result, pure disaster.

Later that year in Japan a new bamboo world unfolded. Unlike the heavy, impenetrable clumps of the Philippines, here were slender stalks in groves that one could walk through. On all sides were bamboo products as pleasing to the eye as they were useful. I would spend hours in village bamboo shops watching how the craftsman used his teeth and toes, as well as his hands, to split culms into fine strands for weaving.

Bamboo rakes and brooms, as well as bamboo poles for drying clothes and airing bedding, were used in every yard. Bamboo fences in classic styles marked property lines. Bamboo water dippers were used at the well. The housewife brought home meat wrapped in a bamboo sheath and prepared bamboo shoots for the table. Her husband smoked a bamboo pipe and walked with a bamboo cane. Her children played with bamboo toys or flew bamboo-ribbed kites. From a neighbor's yard came the sound of a bamboo flute. Bamboo surely was the most useful plant in the world, and I set out to find as many examples as possible of those uses. Because of the sheer abundance of bamboo artifacts, however, I soon had to qualify that goal. I decided to look for pieces that were beautiful, or unusual, or old.

In those early occupation days there was a phrase that applied to many Japanese people: *Take no ko seikatsu*—"living like a bamboo shoot." It compared the forced shedding of one's personal possessions with the shedding, first on one side and then on the other, of the sheaths of the growing bamboo plant. These were difficult days for the formerly rich of Japan. Precious items that might have been in a family for generations were quietly pushed into antique shops for whatever price they would bring. It was my good fortune to be on the scene at that time.

In my early collecting I was enthusiastically joined by my wife, Mona Miwako, whom I met on Christmas Day 1945 in Oiso, a suburb of Tokyo. Together we struck out each free afternoon and weekend for the antique shops of the Kansai area. In Osaka, Kyoto, Nara, and Kobe we became well acquainted with craftsmen and dealers and with bamboo artists like Shōkosai IV and Chikuunsai II, whose works are on display in this exhibition (135 and 152).

We lived in Kobe for one year while I was a member of the Hyogo Prefecture Military Government team. Then in July 1948, we moved to the United States, where for 17 years we engaged in a bamboo specialty business. During that time I was in close contact with Robert A. Young and Dr. Floyd McClure, bamboo specialists with the Department of Agriculture in Washington, D.C. Under Young's skillful guidance we planted cuttings of his bamboo at our home on the shores of Lake Erie and saw amazing growth even in this inhospitable climate. Strangely, both Young and McClure were fellow Ohioans and graduates of Ohio State University. I treasured their friendship and now, after their passing, salute them for the dignity and scholarship they brought to the study of bamboo.

During the Cleveland years, Young was a frequent visitor at our house and a tireless correspondent about things bamboo. He borrowed several of our pieces for the department's botanical study in Washington, one being Shōkosai IV's cabinet with the double-branching handle, a botanical abnormality that he had previously deemed impossible (135).

Although I learned much about bamboo through books, periodicals, letters, and business associations, I was troubled by not being able to add appreciably to our collection. A few objects came by mail from dealer friends in Kobe and Osaka, but only enough to whet my appetite. When in 1965 I was invited to join a Tokyo firm supplying various and sundry goods to the U.S. military throughout the Far East, I gladly accepted. Soon Mona, our two high-school-age daughters, Bonny and Tina, and I were back in Japan. A second phase of our collecting began—now, in truth, a family enterprise.

For me, a happier set of circumstances could not have been desired. Visiting U.S. bases entailed crossing Japan from north to south and from east to west, with occasional desk time and weekends being spent in antique-rich Tokyo. Outside Japan, calls frequently had to be made in Taipei, Seoul, Hong Kong, Manila, Bangkok, and Saigon — all cities where bamboo was much in use. As in earlier years, my free time was spent digging through antique shops. Especially fruitful was Hong Kong with its flow of old bamboo artifacts coming down from Peking, Shanghai, and Canton.

I must confess that my sorties did not always leave me with complete peace of mind. I was often anguished with doubt at the propriety of buying this piece or that, especially when the prices climbed out of the fun-and-games area. Perhaps every collector suffers some such misgivings. Knowing, however, that my worst qualms resulted from failure to buy, I continued on a course of acquisition the full twelve years we were in Japan.

A man should be grateful for his family's patience with such ventures, and I was. What I was not prepared for was the extent of their participation. A wealth of fine pieces was added to the collection through their efforts. Mona claimed many exquisite baskets and accessories at the July and December art shows in Tokyo (154 and 156). Bonny brought home the superb Zeshin *inrō* decorated with newly hatched chick and egg (185). Tina, now married, wrested from the De Menasce collection in London an intricately carved box (23). And Tina's husband, Michael Chow, scratched by the same needle, contributed woven bamboo containers from Africa, canes from London, and choice carvings from Hong Kong (25 and 59). I warmly welcomed all these pieces for what they were — examples of fine bamboo craftsmanship from far-flung places — but I welcomed them even more for their implied endorsement of my course of action.

In 1975, I wrote an article on bamboo brushpots which appeared in the September-October issue of *Arts of Asia*. The article was seen by Mr. and Mrs. David Touff of Denver, who brought the collection to the attention of Curator of Asian Art Ronald Otsuka and Director Thomas Maytham of the Denver Art Museum. After a series of meetings with the parties involved, I accepted the museum's proposal to house the collection in the Asian Art Department. The bulk of the collection was shipped to the museum in October 1977, the first step toward a series of exhibitions in which we are very happy to participate.

If, after viewing this exhibition and perusing this catalog, you are tempted to begin collecting bamboo objects, "Bravo!" But if you collect them, know bamboo. You must be able to recognize what is bamboo and what is not. Dealers are not infallible, nor are they always disinterested. The key to recognition is the grain, both the cross section and the vertical. Study this, and you won't often be fooled.

My wife and I are sincerely grateful to the Touffs, Ronald Otsuka, and Thomas Maytham for their part in making possible the present exhibition. Also most helpful were L. Anthony Wright and Susan Mundt of the Registrar's Office, Margaret Ritchie of the Publications Department, and Bj Averitt, volunteer staff aide in the Asian Art Department. The task of cataloging the many pieces of the collection in the short time allotted was difficult and laborious, but the concerted efforts of these skilled, energetic, and unselfish people resulted in a job beautifully done. We also wish to give special thanks to Lloyd Rule and his staff for producing so well that most essential element of the catalog, the photographs.

Walter E. Lutz

Acknowledgments

My first meeting with Mona and Walter Lutz was on June 12, 1976, when they visited Denver at the invitation of Terry and David Touff. The Touffs, long-time friends of the museum, had read an article about the Lutz Bamboo Collection in which Walter said that he would " . . . welcome from readers of *Arts of Asia* any suggestion as to how and where best to exhibit permanently his pieces, not excepting the possibility of giving the collection to a municipality or institution." In response to this query, the Touffs suggested Denver and the Denver Art Museum as a potential home for the collection and arranged to introduce the Lutzes to Director Thomas Maytham and me.

Half a year later, I saw the Lutzes again while I was in Japan. Traveling from their home in Hiratsuka to meet me in Tokyo, Mona and Walter brought a suitcase packed with wonderful treasures, all made of bamboo. I marveled over the beauty and variety of these bamboo objects. They entrusted to my care one of their prized brushpots (42), which I brought back to Denver as an example of the quality of the Lutz Bamboo Collection.

With the administrative support of Thomas Maytham, Associate Director Lewis Story, and Development Director Cathey Finlon, I spent several months in correspondence with the Lutzes. Finally, an agreement was made to bring the Lutz Bamboo Collection to Denver for a series of exhibitions. In 1977, L. Anthony Wright, the museum's registrar, expertly arranged the shipment, insurance, and customs clearance of over 2,000 bamboo objects from Japan to Denver. He and his assistant, Susan Mundt, then documented and registered these pieces into the museum's collection as long-term loans with the help of Bj Averitt, my volunteer staff aide, and Mona and Walter, who were now living in the United States.

The Lutzes came to Denver in March of this year, and together we reviewed the selection of objects for this exhibition. Mona and Walter shared their intimate knowledge of bamboo with me and generously provided information which they had acquired through years of collecting. Their explanations and identifications were indispensable to the writing of this catalog.

Marlene Chambers and her editorial staff supervised the preparation of this publication. Margaret Ritchie read my manuscript with sensitivity and care, and her suggestions have lent consistency and clarity to the text. The beautiful layout of the catalog was designed by Amelia Ives.

Photographing bamboo objects is unbelievably difficult, and the skill, talent, and painstaking work of Lloyd Rule has produced marvelous results. He and his assistant, Richard Baume, worked diligently and patiently with the aid of their fine volunteers. Laboring far beyond the call of duty, Bj Averitt devoted countless days to coordinating and scheduling the photographic work. She is a rival to bamboo in her flexibility and endurance.

Through his sensitive judgment and esthetic awareness, Exhibition Designer Jeremy Hillhouse has achieved a harmonious presentation of these objects, which were expertly mounted by the museum's installation team. Emma Bunker, Lou Graham, Lorraine Kaimal, Esther Ruckman, and the Asian Art Department docents were most helpful in the preparation and staffing of this exhibition.

I am deeply grateful to Mona and Walter Lutz. Their love of bamboo is the foundation of this project. They have shared their collection warmly and openly in this attempt to further the understanding and appreciation of a most remarkable plant and the artistry it has inspired. My sincerest gratitude is also extended to the many friends whose work, support, and patience have made this exhibition a reality.

R.Y.O.

229

Bamboo

In Asia, bamboo has long been recognized as a source of beauty and as a material of myriad uses. Artists and craftsmen have been inspired by its elegance and simplicity. Poets and philosophers have celebrated its strength and resilience. It provides food, shelter, and tools — the necessities for life and livelihood. Standing straight and withstanding adversity, bamboo possesses qualities that mankind has long endeavored to emulate. It is a symbol of spiritual and moral ideals, an embodiment of respected virtues.

Whether growing in natural groves or in cultivated gardens, bamboo has an inherent beauty. Delicate plumage atop sturdy stems creates an image of propriety and grace. Its leaves glisten in the sunlight; its long, jointed shafts sway in the wind. During peaceful moments, bamboo whispers and shimmers in humble simplicity. Rooted firmly, it can weather great storms, bending without breaking and remaining green throughout the winter.

Bamboo is native to Asia and flourishes in tropical and subtropical zones of warm temperatures and high humidity. It grows particularly well in Japan, China, India, and Southeast Asia. Fossil remains suggest that bamboo existed tens of thousands of years ago. Numerous varieties of bamboo have been currently identified, diverse in their size, coloration, and growth pattern.[1] Apparently, the name *bamboo* is taken from a Malay word imitating the explosive sound the plant makes when it is burned. Marco Polo, in fact, had observed travelers placing several green bamboo poles near a campfire. As the bamboo periodically exploded during the night, marauding animals would be frightened away.[2]

Bamboo is neither a grass nor a tree and has been accorded a botanic family of its own, *Bambusaceae.* Its hollow stem, or culm, which grows rapidly to its full height, is jointed at nodes marked internally by strong diaphragms (figs. a and b). Bamboo follows two main patterns of growth, both determined by the nature of its underground stem, or rhizome, from which new sprouts are generated. The sympodial, or "clump," system of growth (fig. c) is prevalent in the tropics while the monopodial, or "running," type (fig. d) is usually found in cooler climates.

47

1

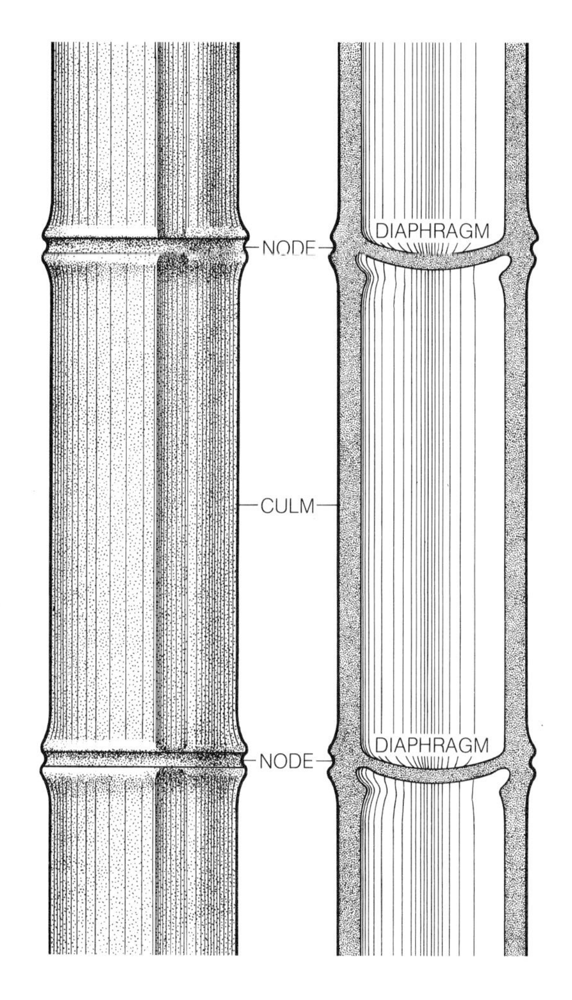

a. Outer view and vertical section of culm

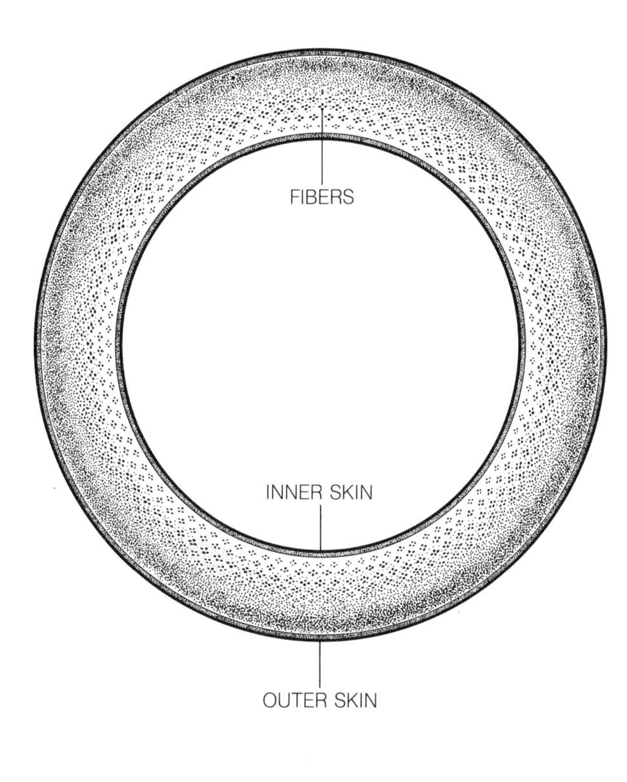

b. Cross section of culm

In the case of the sympodial system, the rhizome puts forth a new shoot which is connected directly to the parent plant. The underground shoot turns upward and forms a new culm, a process which is repeated in short, successive links. The result is a dense clump of bamboo, much less diffuse than the monopodial type, which sends out long underground runners sometimes over a hundred meters in length. Buds are produced along the joints of the horizontal runners, and some of them germinate to emerge above ground as new culms. The rhizomes have a minor system of true roots.[3]

Certainly one of the most versatile of plants, bamboo is made into products serving the most refined purposes, as well as commonplace needs. Uses have been devised for every part of the bamboo plant, but artists and craftsmen have worked primarily with its culm and rhizome. Factors of form, proportion, color, texture, grain, and weight are qualities which define the final product.

In general, the working of bamboo has involved two basic techniques: carving its culm and rhizome (43 and 17), or weaving split strands into baskets, containers, and other articles (150 and 100). In addition, lengths of bamboo have been used for utilitarian objects, such as canes, and in the construction of furniture (213 and 220). Strips of bamboo are sometimes applied to other materials as a decorative veneer, a technique finding special favor in Korea (70). Ranging in color from yellow to red, brown, and almost black, bamboo is usually left in its natural tone. The possibilities of decorating bamboo with lacquer, however, have certainly found success, particularly in Japan (71 and 185).

When bamboo is carved, the tubular culm is most suitable for fashioning brushpots, incense stick containers, wristrests, flower containers, and other curved or cylindrical objects (47 and 157). Particular advantage is taken of the node and diaphragm, which provide an organically integrated base for vessels and containers.[4] By making clever use of the natural septum, trays, dishes, and cups have also been made (127 and 144). In a technique known as "yellow skin" carving, the layer of skin from the interior of the culm is carefully removed and laminated onto a flat surface in preparation for engraving (168 and 169).[5]

Bamboo carvings popularly referred to as "root" carvings are made from bamboo's nearly solid, underground shoot. Figures, animals, miniature landscapes, brushrests, seals, *netsuke,* and a variety of other objects are carved from the shoot, which is frequently inverted before carving to provide a wider base (4 and 18). Its gnarled protrusions are often incorporated into the carved image, and on occasion it has been used for utilitarian items such as pipes and containers (229 and 162).

One of the oldest and most extensive uses of bamboo has been in making woven objects, such as mats, baskets, and containers (98 and 211). Although the earliest woven articles themselves have not survived, impressions left on prehistoric earthenware jars attest to their existence thousands of years ago. In preparation for weaving, the bamboo culm is carefully stripped and split into long segments. The strong outer rind is able to withstand heavy use while the inner culm, which can be split to the diameter of a fine thread, is fairly perishable. Certain reedy types of bamboo have pliable young culms which can be woven without splitting.[6] Many other materials have been used in combination with bamboo to create woven articles (99), and a woven bamboo surface is often lacquered or reinforced with a lacquered lining (95).

Whether it was to be carved or woven, traditional methods for selecting and seasoning bamboo were followed. Depending on the product, the process varied. Certain varieties of bamboo were favored, and the age and size of the plant, as well as the time of year it was uprooted, were primary considerations. The bamboo was seasoned by boiling or heating it to remove oils and chlorophyll, by drying it in the sun or shade at an appropriate temperature and humidity, and sometimes by smoking it. The seasoning procedure often required several years.[7]

Probably the earliest example of worked bamboo that has been discovered is a neolithic Chinese spearhead dating about 6,000 BC. Inscribed bamboo slips, as well as utilitarian bamboo objects, such as bows, baskets, mattresses, boxes, writing brush handles, and a crossbeam from a balance, have been recovered from Chinese tombs of the Warring States period (481-221 BC). From the Han dynasty (206 BC-AD 220), archaeologists in China have found lacquered bamboo wares, fans, ladles, and hairpins.[8] A number of Han dynasty lacquered boxes, bowls, and cups were discovered in tombs around Pyongyang in North Korea. These finds included the famous "painted basket," a rectangular box of split bamboo recovered from the tomb of a minor official of the ancient Chinese colony at Lo-lang.[9]

134

71

154

In the *Book of Southern Ch'i,* reference is made to a *ju-i* scepter, carved from a bamboo root and presented to a virtuous recluse by the Emperor Kao (479-482) of the Ch'i dynasty.[10] The use of bamboo during the T'ang dynasty (618-906) is best illustrated in objects which have survived not in China, but in the Shōsōin repository at Nara, Japan. Musical instruments, writing brushes, arrows, and a covered box with a woven bamboo exterior are among the pieces beautifully preserved there since the middle of the 8th century. A sutra cover in the Shōsōin collection is made of split bamboo which has been interwoven with two colors of silk thread to create an intricate pattern.[11]

The 11th-century history *Experiences in Painting* by Kuo Jo-hsü refers to the achievement of a bamboo carver who created a brushhandle belonging to Wang I, the T'ang dynasty imperial commissioner for Te-chou (Shantung). The brush, which had a thicker shaft than usual, was carved so that

> all the handle from a half inch or so in at the ends was engraved with scenes of "Marching along with the Army." Men, horses—down to their very hair—kiosks, terraces, and far-distant waterways, all were done with the utmost finesse.[12]

By the Sung dynasty (960-1279), bamboo was being carved with a wide range of pictorial representations. The following observation is made by T'ao Tsung-i in *Chou Keng-lu,* a journal of the Yüan dynasty (1260-1368):

> During the reign of Sung Kao-tsung, there was a craftsman named Chan Ch'eng whose ability as a carver remained unchallenged. He once made a bird cage whose four sides were composed of carved bamboo slips. All the palaces, figures, landscape scenes, trees, flowers, birds, and beasts (carved on these slips) were of the most exquisite fineness and vitality. For more than two hundred years no one has been able to surpass his technique.[13]

Although techniques for carving bamboo had been developed earlier, it was during the Ming dynasty (1368-1644) that bamboo carving became a major form of artistic expression. It attracted the attention of late Ming scholars who regarded it as a means of transmitting feelings through the knife as painting and calligraphy did through the brush. With the interest of the literati class, bamboo articles were made specifically for the scholar's studio. Literary subjects, *lohans,* Taoist immortals, legendary figures, auspicious birds, animals, and plants were commonly represented (44 and 15).

As with other art forms subjected to literati connoisseurship, distinctions based on ideological or stylistic differences arose between various schools of bamboo carving. Two centers became predominant, Chia-ting and Chin-ling; both were located in Kiangsu province. Members of the Chia-ting school advocated high relief carving with deep cutting, robust forms, and occasional openwork. They were headed by three generations of the Chu family: Chu Sung-lin, his son Chu Hsiao-sung, and his grandson Chu San-sung.[14]

In contrast, the Chin-ling school, centering around Nanking, was noted for the simplicity, shallowness, and spare quality of its bamboo carvings. It maintained that this light and delicate manner of carving befitted a scholar. Excessive attention to detail was eschewed as the concern of artisans. *A Record of Things Heard and Seen in the Ch'u-mu Pavilion* includes the following comment about P'u Chung-ch'ien, the head of the school:

> When he put his knife to bamboo, his skill rivalled that of the immortals. Sometimes he would merely scratch a few lines on a piece of bamboo and the result would far excel anything other carvers of the day could accomplish.[15]

A famous master of the 17th century who did not belong to either of these schools was Chang Hsi-huang of Chia-hsing in Chekiang province. He is best known for his use of a bamboo carving technique called *liu-ch'ing yang-wen,* meaning "leave the skin in relief." By this process, the light-colored surface of the bamboo skin was left in reserve. When carved through to the darker layers below, an effect similar to shading was achieved.[16]

The scholarly quality of Ming bamboo carving continued into the beginning of the Ch'ing dynasty (1644-1912). Earlier techniques were modified, or several different techniques were incorporated in a single piece. The antiquarian interest of the Ch'ien-lung period (1736-1796) brought about a vogue for archaistic forms recalling ancient bronze vessels (26 and 28). Chia-ting prospered as the leading center of bamboo carving, but from the 18th century, other districts began producing noteworthy carvers. The carving of fan frames (170 and 171) became independently popular by the end of the Ch'ing dynasty, and various styles and techniques of bamboo carving continue to thrive during the 20th century.[17]

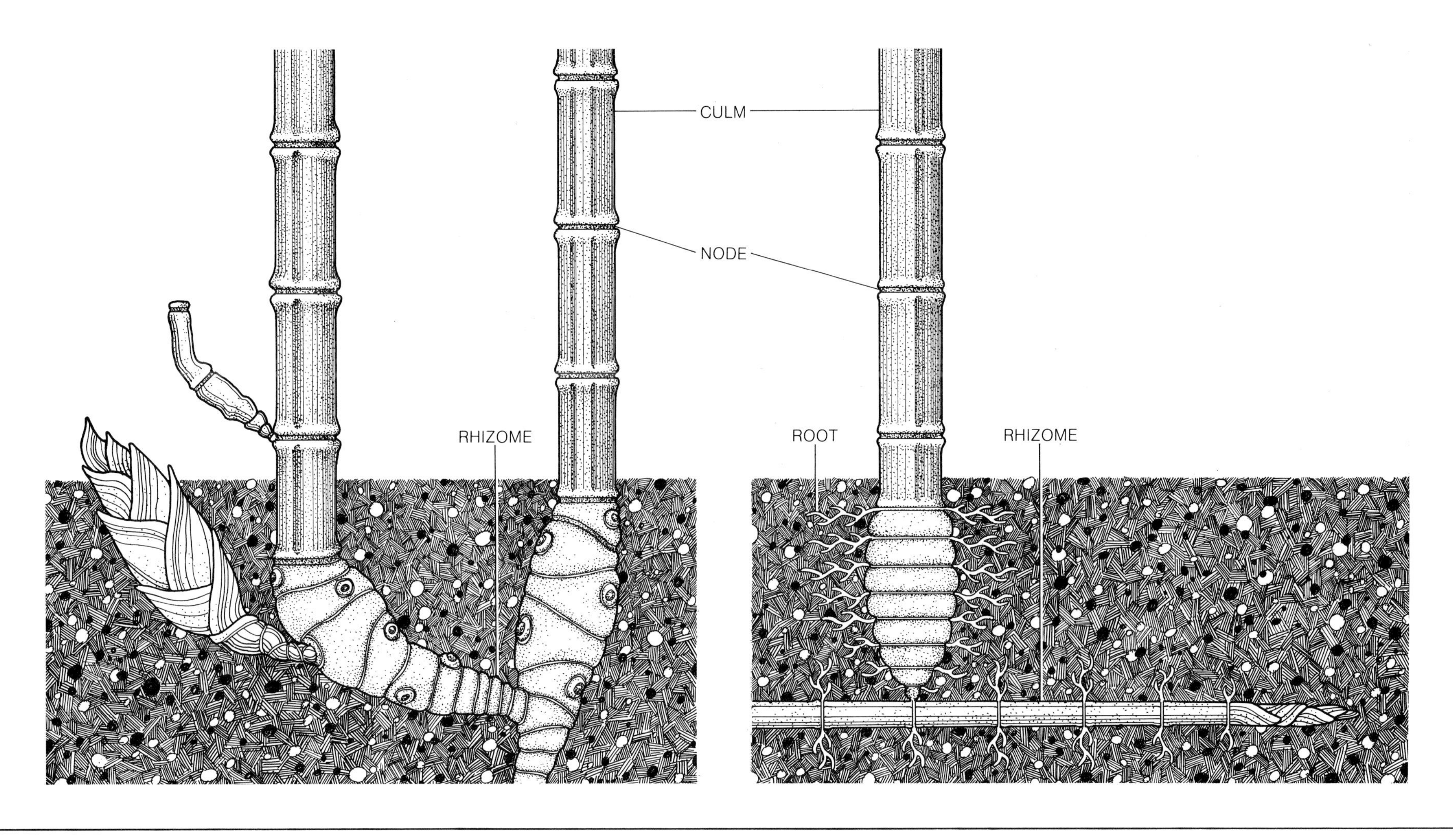

c. Sympodial system of growth

d. Monopodial system of growth

102

185

In addition to carved bamboo, woven baskets have held a prominent position in China since the earliest times. They were used for rituals and ceremonies, the presentation of wedding gifts and funeral offerings, and a variety of industrial purposes from fishery to sericulture. In northern China, baskets were chiefly used for agriculture and were, by necessity, strong and practical. Those from the central and southern parts of the country were often designed for pleasurable pastimes such as picnics (96). Woven bamboo was combined with wood, metal, and lacquer, creating a broad range of decorative possibilities.[18]

In Korea, as in China, bamboo was a favored material for objects destined for the scholar's studio. Unlike the single-culm brushpots of China, Korean examples combine several sections of culm on a wooden base (67 and 68). The exterior surface of the pot was carved and lacquered, making a strong contrast between the raised design and its ground. Bookcases, stationery chests, letter holders, and writing boxes were also frequently constructed in part with bamboo.

Bamboo articles were produced mainly in the Cholla-do area at the southwestern tip of the Korean peninsula.[19] Display shelves, wardrobes, tables, chests, and boxes were decorated with a mosaic pattern of bamboo veneer,[20] and in Chungchong-do, slightly to the north, the bamboo was "branded" with decorative motifs.[21] Utilitarian products such as quivers and paper lantern frames were among the many uses devised for bamboo by Korean craftsmen (175 and 176).

The inexhaustible ingenuity which bamboo inspires is seen in Japan as well. In work or play, indoors or out, bamboo had its place. Music, archery, painting, and writing were among the many activities dependent upon bamboo. With amazing originality, popular subjects were fashioned from bamboo into the three-part objects used by the Japanese to carry personal items: the *inrō,* or "case"; *ojime,* or "bead"; and *netsuke,* or "toggle" (178). Since the traditional Japanese garment is without pockets, the *inrō* was designed to be worn at the sash *(obi).* Originally made to carry the seal and ink essential for transacting business, the *inrō* was later adapted to hold medicines, powders, or small objects (179 and 182).

The quintessence of bamboo's role in Japanese culture is brought into focus by the tea ceremony *(chanoyu).* Essentially the preparation of a cup of tea for a friend, the tea ceremony has been refined to the highest level of esthetic awareness. It pervades Japanese life and is unsurpassed for the manner in which it incorporates various arts into a harmonious relationship. The cultures of China, Korea, and Japan and the talents of aristocrats, peasants, and artists have all made their contributions to *chanoyu.*

The integral role of bamboo in the tea ceremony has been established over the centuries through the sensibility of great tea masters such as Sen no Rikyū (1521-1591). As a natural material, bamboo fits the spirit of *wabi* tea, or "poverty" tea, which was influenced by Zen Buddhism and its renunciation of material things.[22] Bamboo utensils, paintings, calligraphy, ceramics, lacquer, wood, and metalwork, textiles, flower arrangements, gardens, and architecture are selected with great sensitivity to complement one another and the particular spirit and mood of the host, guest, and day. Sharing the occasion together, all of these elements embody the principles of harmony, humility, purity, and peace.

In experiencing the tea ceremony, the initial impression is made by the garden path which leads to the tea house. Swept clean by a bamboo broom, it is generally defined by fences and gates made of green bamboo. Parts of the tea house itself are made of bamboo—plastered latticework walls, grid windows, and coverings for the ceiling and roof. Inside the tea house, attention is focused on the *tokonoma* ("alcove"), which may be framed on one side by a vertical bamboo column. A painting or calligraphy is displayed in this special recess, and a flower arrangement is either placed on the *tokonoma* floor or suspended from its back wall or side column. If the season is appropriate, the flower container may be made of bamboo, either cut in natural segments or woven into a basket (164 and 151).

From the ceiling hangs a bamboo *jizai,* a pole used to support the kettle over the hearth (210). The *jizai* symbolizes a bridge between heaven and earth, the position traditionally held by mankind in Buddhist thought. This symbolism is paralleled by the stand upon which the major tea utensils are arranged.

95

Depending upon the type of tea served and the objects selected for the ceremony, a bamboo stand may be used. It is composed of a rectangular board and upper shelf of wood joined at the corners by four bamboo supports. On the lower *jiita* ("earth board") rest the kettle, its brazier, and the water jar. The *tenita* ("heaven board") above holds the tea caddy and teabowl, objects symbolizing the sun and the moon. It is again bamboo which metaphorically joins heaven and earth. A tea scoop (105), usually made of bamboo, is used to transfer measured amounts of powdered tea from the caddy to the teabowl. Then a water ladle is used to pour hot water into the teabowl, and the tea and water are blended with a tea whisk (109). The water ladle and whisk are two utensils which are invariably made of bamboo. Another such item is the lid rest (110), a simple section of bamboo designed by the great tea master Rikyū.[23] Usually bamboo items such as these are intended to be used only once. However, if created by a tea master or noted person, they have been carefully preserved, acquiring a fine patination through repeated use.

Bamboo is used for other articles associated with the tea ceremony, including incense containers, charcoal baskets, and smoking paraphernalia (142 and 136). The pleasures of dining are yet another aspect of some tea ceremonies, and bamboo trays, dishes, and wrappings for sweets are often used. Also fashioned from bamboo are sticks for handling fish, pickles, and grilled food. Finally, there is the folded paper fan with a bamboo frame which is carried by the guest.[24]

Each of the utensils used in the tea ceremony bears a long tradition of artistic craftsmanship. They are often categorized according to degrees of formality: *shin* ("formal"), *gyō* ("less formal"), and *sō* ("informal"). Even the unassuming tea scoop is differentiated into such classifications. Sometimes a highly regarded utensil is given a name which is inscribed on its storage container, confirming its honored status. For example, Sen no Rikyū himself is said to have originated a type of flower container made from a section of bamboo that is notched at its top. The oldest and most famous container of this type made by Rikyū is called *"Onjōji."*[25] Now in the Tokyo National Museum *Onjōji* became the prototype for similar vessels made through the years (157).

Flower arrangements for the tea ceremony are meant to appear as if still growing in the fields. Flowers are selected to suit the season and the room in which they are displayed.[26] They are generally arranged first in the hand and are then placed in the container, where they should look free and natural without being wild or haphazard. Bamboo flower containers, of course, are frequently used for occasions other than tea ceremonies. These vessels and baskets cover a variety of types and shapes, but they share in common the unique quality of bamboo, which acts as a natural partner to blossoms, buds, and grasses.

Bamboo is joined by the plum and pine trees in a group called the "Three Friends." Bamboo remains green throughout the year; the pine is renowned for its longevity; and the plum, which flowers at the coldest time of the year, is a symbol of beauty and strength. As a group they connote happiness and good fortune (103 and 128). On another level of meaning they represent Sakyamuni, Lao-tzu, and Confucius, the founders of Buddhism, Taoism, and Confucianism (11). In this context, bamboo, plum, and pine are Three Friends symbolically joining these doctrinal schools in harmony and unity.[27]

In nature, bamboo evokes regularity, freshness, and simplicity. Through the imagination of craftsmen, it has been transformed into utilitarian articles of endless variety. In the hands of artists, bamboo is transfigured into objects with a beauty capable of expressing the highest ideals of mankind.

Ronald Y. Otsuka
Curator of Asian Art

26

25

Notes

1. Fujino Junko mentions as many as 600 varieties ("Bamboo," *The East* 8, no. 9 [October 1972]: 20). In a historical summary of bamboo as a material, Ip Yee and Laurence C. S. Tam note that by the 6th century, more than 70 species of bamboo were known in China. During the Yüan dynasty (1260-1368), the number had increased to over 334 species. Today botanists record over 20 genera and more than 1,000 species (*Chinese Bamboo Carving,* Part 1 [Hong Kong: Hong Kong Museum of Art, 1978], p. 21).

2. Robert Austin and Kōichirō Ueda, *Bamboo* (New York and Tokyo: Weatherhill, 1970), p. 10.

3. For a discussion of bamboo and its growth and cultivation, including a description of the phenomenon of bamboo flowering, see Austin and Ueda, *Bamboo,* pp. 192-208. This topic is also treated by Fujino, "Bamboo," pp. 19-27. A comparison of catalog nos. 229 and 162 reveals the difference between sympodial and monopodial rhizomes.

4. Walter E. Lutz describes how brushpots are made from an inverted segment of culm cut about an inch above the node and six inches below it ("Bamboo Brushpots," *Arts of Asia* 5, no. 5 [September-October 1975]: 28).

5. J.W.H. Grice, "Chinese Bamboo Carvings," *Country Life,* 6 May 1954, p. 1403, and Willem Irik, "Chinese Woodcarvings of the Ch'ing Period," *Arts of Asia* 8, no. 5 (September-October 1978): 86.

6. Barbara Adachi, *The Living Treasures of Japan* (Tokyo, New York, and San Francisco: Kodansha International, Ltd., 1973), pp. 57-58, and Victor Hauge and Takako Hauge, *Folk Traditions in Japanese Art* (International Exhibitions Foundation, 1978), p. 31.

7. Ip and Tam, *Chinese Bamboo Carvings,* p. 23. Adachi provides an interview with bamboo artist Shōunsai Shōno, who describes his methods for seasoning bamboo (*Living Treasures,* p. 57).

8. Ip and Tam give a summary of bamboo objects recently excavated in China with references to archaeological reports published in the Chinese journals *Kaogu* and *Wen Wu* (*Chinese Bamboo Carving,* pp. 15-16).

9. Laurence Sickman and Alexander Soper, *The Art and Architecture of China* (Baltimore, Maryland: Penguin Books, 1956), p. 34, and Evelyn McCune, *The Arts of Korea: An Illustrated History* (Rutland, Vermont and Tokyo: Charles E. Tuttle Company, 1962), pp. 44-45.

10. Ip and Tam, *Chinese Bamboo Carvings,* p. 16. A later example of a *ju-i* scepter is represented by catalog no. 29.

11. *Treasures of the Shōsōin* (Tokyo: Asahi Shimbun Publishing Co., 1965), pls. 35, 36, and 48, and Kiuchi Takeo, *Bokuchiku Kōgei* [Wood and Bamboo Crafts], Nihon no Bijutsu [Arts of Japan], no. 25 (Tokyo: Shibundō, 1968), pls. 3 and 111.

12. Alexander Coburn Soper, trans., *Kuo Jo-Hsü's Experiences in Painting (T'u-hua Chien-wên Chih): An Eleventh Century History of Chinese Painting together with the Chinese Text in Facsimile* (Washington, D.C.: American Council of Learned Societies, 1951), pp. 85-86.

13. Na Chih-liang, "Some Bamboo Carving Techniques of the Ming Dynasty," *The National Palace Museum Bulletin* 2, no. 3 (July 1967): 7.

14. Ip and Tam, *Chinese Bamboo Carving,* p. 17, and Na Chih-liang, "Wood and Bamboo Carvings in the National Palace Museum, Taipei," *Arts of Asia* 7, no. 4 (July-August 1977): 61-62. An example of this style is represented by catalog no. 41. Ip and Tam, pp. 24-25, note that since bamboo carving was part of a traditional scholarly way of life, it lacked the convenient reign marks of the court-dominated art forms. Thus, the precise dating of bamboo objects is difficult.

15. Na, "Wood and Bamboo Carvings," pp. 61-62. Members of the Chia-ting school considered work of the Chin-ling school to be shallow and hastily executed.

16. Ip and Tam, *Chinese Bamboo Carving,* p. 18, and Na, "Wood and Bamboo Carvings," p. 62. For an example of this technique, see catalog no. 46.

17. Ip and Tam, *Chinese Bamboo Carving,* pp. 19-20.

18. Berthold Laufer, *Chinese Baskets,* Anthropology Design Series, no. 3 (Chicago: Field Museum of Natural History, 1925), pp. 3-4, and Tseng Yu-ho Ecke, *Chinese Folk Art II: In American Collections, from Early 15th Century to Early 20th Century* (Honolulu, 1977), pp. 3 and 8.

19. Kim Wan-gyu, *Mok'chil Kongye* [Technology of Wooden Lacquerwork], Vol. 2 of *T'ongin misul* [Tong-in Arts] (Seoul: Tasangak, 1975), no pagination.

20. Choi Sun-wu and Chŏng Yang-mo, comps., *Mok'chil Kongye* [Technology of Wooden Lacquerwork], Han'guk misul chŏnjip [Collections of Korean Art], no. 13 (Seoul: Tonghwa Chulpan Kongsa, 1974), pls. 13, 19, 21, 38, 44, 60, and 81.

21. Kim, *Mok'chil Kongye,* pl. 70.

22. Hayashiya Seizō, *Chanoyu: Japanese Tea Ceremony,* trans. Emily J. Sano (New York: Japan Society, 1979), p. 18.

23. "Chadogu—Tea Utensils: Bamboo," *Chanoyu,* no. 14 (1976), pp. 54-55 and 58, and Fujioka Ryōichi, *Tea Ceremony Utensils,* trans. Louis Allison Cort, Arts of Japan, no. 3 (New York and Tokyo: Weatherhill and Shibundo, 1973), p. 71.

24. "Chadogu," pp. 56-57 and 59.

25. Hayashiya, *Chanoyu,* p. 95.

26. Hayashiya, *Chanoyu,* p. 87, and Henry Mittwer, *The Art of Chabana: Flowers for the Tea Ceremony* (Rutland, Vermont and Tokyo: Charles E. Tuttle Company, 1974), p. 40.

27. Austin and Ueda, *Bamboo,* p. 21, and Jan Fontein and Money L. Hickman, *Zen Painting and Calligraphy* (Boston: Museum of Fine Arts, 1970), p. 92.

3

Selected Bibliography

Adachi, Barbara. *The Living Treasures of Japan.* Tokyo, New York, and San Francisco: Kodansha International, Ltd., 1973.

Austin, Robert and Ueda, Kōichirō. *Bamboo.* New York and Tokyo: Weatherhill, 1970.

Boode, Peter. "Chinese Writing Accessories." *Oriental Art* 1, no. 3 (Winter 1948): 124-128.

"Chadogu—Tea Utensils: Bamboo." *Chanoyu,* no. 14, (1976), pp. 54-59.

Choi, Sun-wu and Chŏng, Yang-mo, comps. *Mok'chil Kongye* [Technology of Wooden Lacquerwork]. Han'guk misul chŏnjip [Collections of Korean Art], no. 13. Seoul: Tonghwa Chulpan Kongsa, 1974.

Ecke, Tseng Yu-ho. *Chinese Folk Art II: In American Collections, from Early 15th Century to Early 20th Century.* Honolulu, 1977.

Edmund, Walter. "Collector's Column." *Arts of Asia* 7, no. 2 (March-April 1977): 90-92.

Feddersen, Martin. *Japanese Decorative Art: A Handbook for Collectors and Connoisseurs.* Translated by Katherine Watson. New York: Thomas Yoseloff, 1962.

Fontein, Jan, and Hickman, Money L. *Zen Painting and Calligraphy.* Boston: Museum of Fine Arts, 1970.

Fujino, Junko. "Bamboo." *The East* 8, no. 9 (October 1972): 19-27.

Fujioka, Ryōichi. *Tea Ceremony Utensils.* Translated by Louise Allison Cort. Arts of Japan, no. 3. New York and Tokyo: Weatherhill and Shibundō, 1973.

Grice, J.W.H. "Chinese Bamboo Carvings." *Country Life,* 6 May 1954, pp. 1402-1404.

Hauge, Victor, and Hauge, Takako. *Folk Traditions in Japanese Art.* International Exhibitions Foundation, 1978.

Hayashiya, Seizō. *Chanoyu: Japanese Tea Ceremony.* Translated by Emily J. Sano. New York: Japan Society, 1979.

Ip, Yee, and Tam, Laurence C.S. *Chinese Bamboo Carving,* Part 1. Hong Kong: Hong Kong Museum of Art, 1978.

Irik, Willem. "Chinese Woodcarvings of the Ch'ing Period." *Arts of Asia* 8, no. 5 (September-October 1978): 80-90.

Kim, Wan-gyu. *Mok'chil Kongye* [Technology of Wooden Lacquerwork]. Vol. 2 of *T'ongin misul* [Tong-in Arts]. Seoul: Tasangak, 1975.

Kiuchi, Takeo. *Bokuchiku Kōgei* [Wood and Bamboo Crafts]. Nihon no Bijutsu [Arts of Japan], no. 25. Tokyo: Shibundo, 1968.

Laufer, Berthold. *Chinese Baskets.* Anthropology Design Series, no. 3. Chicago: Field Museum of Natural History, 1925.

Luard, Tim. "Chinese Bamboo Carving: A Hong Kong Exhibition." *Arts of Asia* 9, no. 1 (January-February 1979): 52-55.

Lutz, Walter E. "Bamboo Brushpots." *Arts of Asia* 5, no. 5 (September-October 1975): 23-33.

Masterpieces of Chinese Miniature Crafts in the National Palace Museum. Taipei: National Palace Museum, 1971.

Masterpieces of Chinese Writing Materials in the National Palace Museum. Taipei: National Palace Museum, 1971.

McCune, Evelyn. *The Arts of Korea: An Illustrated History.* Rutland, Vermont and Tokyo: Charles E. Tuttle Company, 1962.

Mittwer, Henry. *The Art of Chabana: Flowers for the Tea Ceremony.* Rutland, Vermont and Tokyo: Charles E. Tuttle Company, 1974.

Na, Chih-liang. "Some Bamboo Carving Techniques of the Ming Dynasty." *The National Palace Museum Bulletin* 2, no. 3 (July 1967): 3-7.

__________. "Wood and Bamboo Carvings in the National Palace Museum, Taipei." *Arts of Asia* 7, no. 4 (July-August 1977): 58-62.

Peplow, Evelyn S. "A Heritage of Bamboo Carvings." *Orientations* 10, no. 1 (January 1979): 33-41.

Roberts, Laurance P. *A Dictionary of Japanese Artists: Painting, Sculpture, Ceramics, Prints, Lacquer.* Tokyo and New York: Weatherhill, 1976.

Sickman, Laurence and Soper, Alexander. *The Art and Architecture of China.* Baltimore, Maryland: Penguin Books, 1956.

Soper, Alexander Coburn, trans. *Kuo Jo-Hsü's Experiences in Painting (T'u-hua Chien-wên Chih): An Eleventh Century History of Chinese Painting together with the Chinese Text in Facsimile.* Washington, D.C.: American Council of Learned Societies, 1951.

Treasures of the Shōsōin. Tokyo: Asahi Shimbun Publishing Co., 1965.

6

8

Catalog

The medium for all objects is bamboo unless otherwise stated.

•Objects illustrated in catalog

Carvings and Figures

•1
Chung K'uei with two demons
signed: [Chu] San-sung
China, 17th century, 15.5 cm. h.
The legendary demon queller is shown with a sword and military boots. A demon playing a flute sits beside him while another climbs upon his shoulder and reaches into his ear.
color plate p. 9

2
Lohan
signed: [Chu] San-sung
China, 17th century, 6.5 cm. h.
A *lohan* (Sanskrit, *arhat*; Japanese, *rakan*) is a Buddhist sage who has achieved enlightenment. Usually depicted as a monk or ascetic, he possesses supernatural powers achieved through religious austerities.

10

9

11

•3
Lohan with dragon
China, 18th century, 6.6 cm. h.
With his alms bowl on a rock behind him, the *lohan* holds a pearl. A dragon's head emerges out of the water below him. In Sanskrit this *lohan's* name is Panthaka.
illustrated p. 22

•4
Shou Lao
China, 17th century, 13.5 cm. h.
Holding a peach and a *ju-i* scepter, the bearded god of longevity is depicted with his characteristic elongated forehead. *illustrated p. 3*

5
Liu Hai with the three-legged toad
China, 18th century, 9 cm. h.
Liu Hai is represented holding a magical pill in his left hand and patting the three-legged toad with his right.

•6
Liu Hai with the three-legged toad
China, late 18th century, 24 cm. h.
Seated upon the toad, Liu Hai is shown about to give the three-legged creature a magical pill.

7
Immortal with bats in a basket
China, 18th century, 9.5 cm. h.
The words for *bat* and *good fortune* in Chinese are homophones pronounced "fu." Therefore, bats are often shown as symbols of happiness.

•8
Immortals with bats emerging from a gourd
China, late 18th century, 26.3 cm. h.
In Taoist lore, a gourd is frequently represented as a bottle from which clouds and animal forms emerge.

12

13

•9
Ho-Ho Immortals
China, late 18th century, 5 cm. h.
This pair of figures, one under a lotus flower and the other with a scepter, symbolizes harmonious union.

•10
Li T'ieh-kuai
China, 18th century, 14.5 cm. h.
The immortal, Li T'ieh-kuai, a pupil of Lao-tzu, could send his soul away from his body. When the soul returned and found its own body decomposing, it entered the corpse of a recently deceased lame beggar.

•11
The Three Friends
China, 19th century, 4 cm. h.
This miniature carving represents Sakyamuni, Lao-tzu, and Confucius, the founders of Buddhism, Taoism, and Confucianism.

•12
Two old men smoking pipes
China, 19th century, 21.6 cm. h.
Published: Ip and Tam, *Chinese Bamboo Carving,* #176, and Peplow, "A Heritage of Bamboo Carvings," cover.

•13
Boys with buffaloes
China, 19th century, 21 cm. h.
The three boys and two buffaloes have been carved from the culm of a bamboo plant rather than the usual rhizome. Therefore, the interior is hollow.
Published: Ip and Tam, *Chinese Bamboo Carving, #191.*

14

•14
Boy with buffalo
China, 19th century, 31.8 cm. l.
The boy and buffalo are carved separately. In this case even the stand has been made of bamboo.
Published: Ip and Tam, *Chinese Bamboo Carving,* #192.

•15
Log boat with The Eight Immortals crossing the sea
China, 19th century, 15.2 x 29.9 cm.
The Eight Immortals were invited to share the peaches of immortality at the birthday celebration of the Queen Mother of the West. While on the way to the party, they transformed a tree stump into a boat and crossed the Eastern Sea.
Published: Ip and Tam, *Chinese Bamboo Carving,* #174, and Luard, "Chinese Bamboo Carving," p. 52.
color plate p. 4

16
Log boat with immortals celebrating the birthday of the Queen Mother of the West
China, 19th century, 27 x 34.3 cm.
Published: Ip and Tam, *Chinese Bamboo Carving, #175.*

17

18

20

•17

Miniature mountain with immortals at a birthday celebration for the Queen Mother of the West

China, 19th century, 31.3 cm. h.

The Queen Mother of the West descends from the clouds riding on a phoenix. Standing on a terrace beneath a peach tree, several immortals await her arrival. A log boat floats in at the base of the mountain.

Published: Ip and Tam, *Chinese Bamboo Carving,* #195.

•18

Miniature mountain with pavilions and figures

China, 19th century, 40 cm. h.

This and the previous mountain are done in the style of the 18th century. They are similar to forms found in jadė carving.

Published: Ip and Tam, *Chinese Bamboo Carving,* #196.

•19

Peaches and bat

China, 18th century, 10.5 cm. h.

In this combination of auspicious symbols, the peaches represent longevity and the bat happiness.

•20

Monkeys on peach mound

China, 19th century, 32.4 cm. h.

Published: Ip and Tam, *Chinese Bamboo Carving,* #178, and Luard, "Chinese Bamboo Carving," p. 53.

21

19

23

•21
Deer and cranes
China, late 19th century, 29.5 cm. h.
The deer and crane are symbols of long life and are frequently shown accompanying immortals.

22
Phoenix
China, 19th century, 37.4 cm. h.
Published: Ip and Tam, *Chinese Bamboo Carving,* #193.

•23
Box with pine trees and squirrels
China, 17th century, 4 x 10.5 cm.

24
Box in the shape of a peach
China, 18th century, 4.5 x 8.5 cm.

•25
Box in the shape of two melons
China, 18th century, 6.2 x 14.5 cm.
The overlapping of the two melons is also indicated inside the box.
illustrated p. 20

•26
Hu
China, Ch'ien-lung period (1736-1796), 21.4 cm. h.
Sections of bamboo have been carved and joined into an archaistic form of the ancient bronze vessel. It is decorated with *t'ao-t'ieh* masks and most likely once had a lid.
illustrated p. 19

28

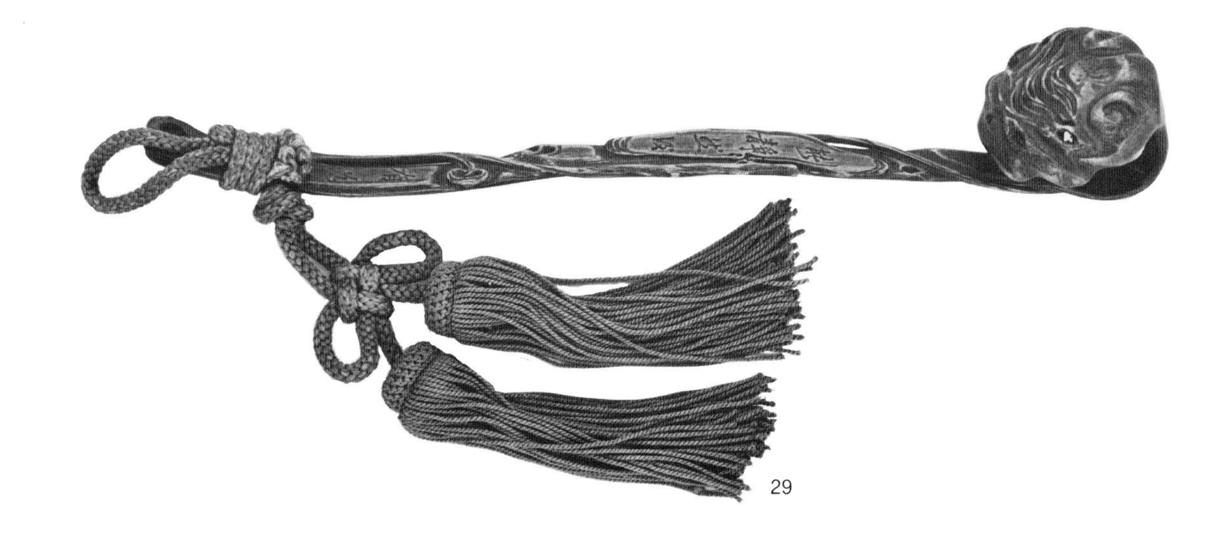

29

27
Ku
China, late 18th century, 16.8 cm. h.
The beaker and its footed stand are carved from the same section of culm. Archaistic motifs of flanges and rising blades are combined with four trigrams.

•28
Ting
China, late 18th century, 20 cm. h.
Carved from a section of culm, the body of this tripod is finely inscribed with the Heart Sutra.

•29
Ju-i scepter
China, 18th century, 29.8 cm. l.
The earliest record of a *ju-i* scepter carved from bamboo appears in the *Book of Southern Ch'i.* It was presented as an imperial gift by the 5th-century Emperor Kao to a virtuous recluse. This scepter is carved from a length of culm which has been bent at its node.

30
Ju-i scepter
China, Ch'ing dynasty (1644-1912), 35 cm. l.

•31
Orchids
Japan, 18th or 19th century, 16.2 cm. h.
This ornamental carving of orchids is pieced together from several sections of bamboo.

31 33 34

32
Cabbage and beetle
Japan, 18th or 19th century, 13.4 cm. l.
The cabbage, which has a small beetle crawling up its leaf, is carved completely in the round.

•33
Snail
Japan, 19th century, 2.4 x 7 cm.

•34
Crab
Japan, 19th century, 16 cm. l.

35
Toad
Japan, 19th century, 4.5 x 8.3 cm.
Effective use is made of the rhizome and its roots to create the textured back of the toad.

36
Pair of deer
Japan, 19th century, 14.3 and 8.2 cm. h.
In Japan the deer, a symbol of long life, is associated with moonlit autumn forests and the Kasuga Shrine in Nara.

37
Hotei (Chinese, *Pu-tai*)
Japan, 19th century, 22.5 cm. h.
Hotei is a nickname meaning "hemp bag" given to a Chinese monk who, in popular Japanese belief, was included among the Seven Gods of Good Fortune.

38

39

•38

Kannon within a shrine

Shimizu Fusae (b. 1910, art name Senshū)
Japan, after 1942. Image 5.6 cm. h., shrine 20.4 cm. h.
In 1942, Shimizu Fusae set herself the task of carving 1,000 Kannons. Using the natural flare of the node as a skirt, she imbues a small piece of bamboo with the spiritual grace befitting the merciful deity. Inscribed with the number 57, this figure stands on a lotus pedestal and is enclosed within a shrine made entirely of bamboo.

•39

Stand

Japan, 19th century, lacquered bamboo and wood, 6.1 x 10.6 cm.

40

Stand

Japan, 19th century, 14.8 x 8.9 cm.

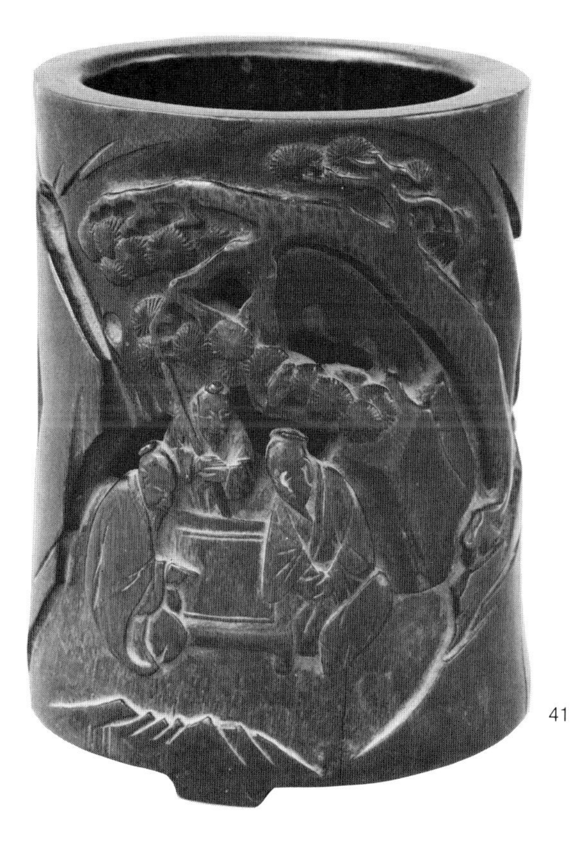

41

42

•41
Brushpot with scholars playing chess under a pine tree
China, 17th century, 14.9 x 11.7 cm.
Published: Lutz, "Bamboo Brushpots," p. 26.

•42
Brushpot with "Poetry Gathering at the Orchid Pavilion"
China, 17th century, 14.3 x 12.4 cm.
On a warm day in March 353, the gentleman-poet Wang Hsi-chih invited 41 of his scholarly friends to the Orchid Pavilion. There they composed poetry and drank wine from cups sent floating down the winding stream. This event became a favorite theme for later generations of literati.
Published: Lutz, "Bamboo Brushpots," p. 27.

•43
Brushpot with the "Seven Sages of the Bamboo Grove"
China, 18th century, 15 x 13.3 cm.
The Seven Sages, who lived during the 3rd century, became famous as lovers of wine and devotees of Taoism. Forsaking strait-laced official life, they retreated to the Bamboo Grove.
Published: Lutz, "Bamboo Brushpots," p. 24.
cover illustration

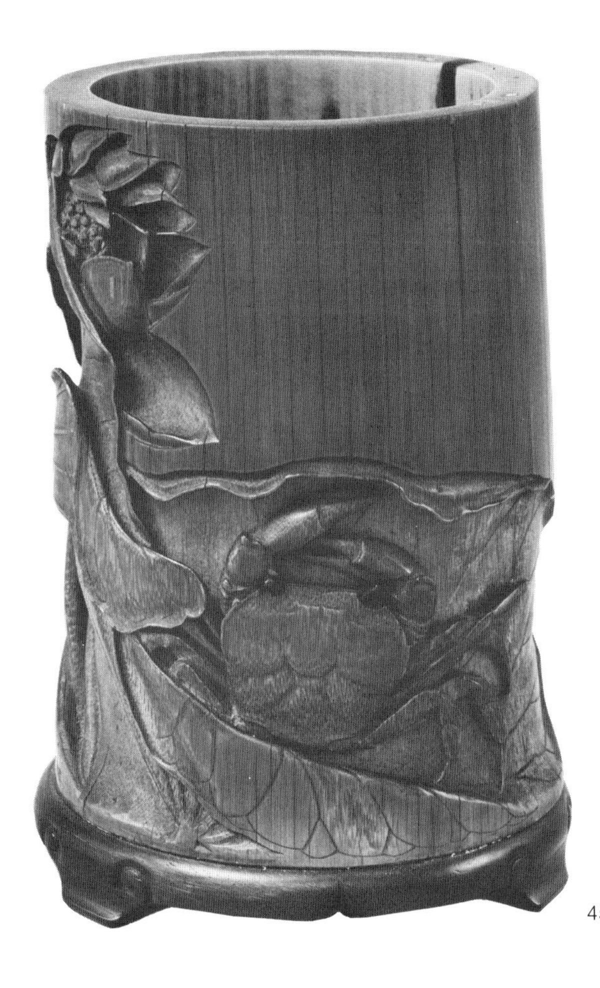

45

46

44
Brushpot with "Evening Cruise under the Red Cliff"
China, 18th century, 14.8 x 11.5 cm.
Su Shih (1037-1101), one of China's greatest poets, composed a prose poem when he visited the Red Cliff, a scenic spot on the Yang-tzu River. It became one of the best-loved poems in Chinese literature and a favorite subject for painters.
Published: Lutz, "Bamboo Brushpots," p. 24.

•45
Brushpot with crab in a lotus pond
China, 19th century, wooden base attached, 17.8 x 12.4 cm.

•46
Wristrest with landscape
China, 18th century, 20.5 x 5 cm.
This is an example of *liu-ch'ing* relief carving. The light-colored outer skin of the culm is left in reserve, contrasting with the darker shades of more deeply carved areas.

•47
Wristrest with horse and pine tree
China, late 17th or early 18th century, 24 x 7.1 cm.
color plate p. 9

48
Wristrest with calligraphy
China, Ch'ing dynasty (1644-1912), 17.7 x 7 cm.

50

51

54

49
Wristrest with calligraphy
China, Ch'ing dynasty (1644-1912), 37.4 x 7.7 cm.
A compressed section of culm was used to make this particular article.

•50
Brushrest with pine trees
China, late 17th or early 18th century, 5.3 x 10.8 cm.

•51
Brushrest with two dragons among clouds
China, late 17th or early 18th century, 3.4 x 11.6 cm.

52
Brushrest with *ling-chih* fungus
China, 18th century, 4.2 x 8.5 cm.

53
Brushrest with two immortals
China, Ch'ing dynasty (1644-1912), wooden base attached, 4.8 x 9.2 cm.

•54
Brushrest with birds on a tree and a basket
China, Ch'ing dynasty (1644-1912), 8.8 x 11 cm.

57

55

•55
Brushwasher in the shape of a pomegranate
signed: Chu San-sung
China, 17th century, 7.2 cm. h.
The brushwasher is inscribed with a 60-year cyclical date. During the late Ming dynasty, it would correspond to the year 1624.

56
Brushwasher with pine branch
China, late 17th or early 18th century, 3.1 x 11.7 cm.

•57
Water container with pine branch
China, 18th century, 3.7 cm. h.

•58
Round seal with lion
China, 17th century or later, 4.2 cm. h.

•59
Square seal with horse
China, Ch'ing dynasty (1644-1912), 5.7 cm. h.

60
Rectangular seal with elephant and two men
China, Ch'ing dynasty (1644-1912), 4.9 cm. h.

58

59

64

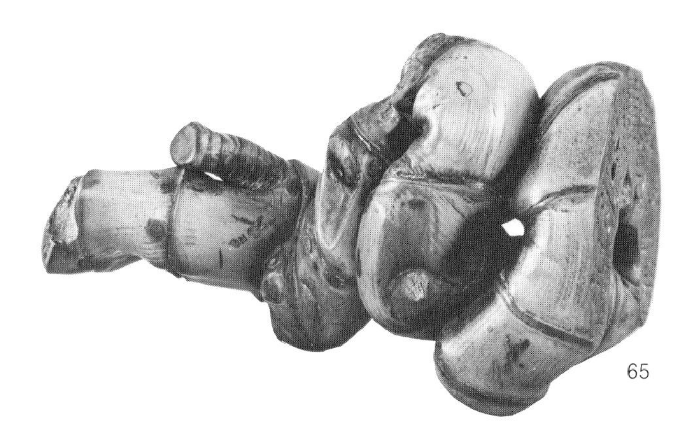
65

61
Rectangular seal
China, Ch'ing dynasty (1644-1912), 4.1 cm. h.
Cut from a section of rhizome, this seal retains the natural roots.

62
Rectangular seal
China, Ch'ing dynasty (1644-1912), 4.6 cm. h.
This seal is made from a cross section taken at the tip of a rhizome.

63
Seal in the shape of a scroll
China, Ch'ing dynasty (1644-1912), 4 cm. h.
Carved to imprint a scroll-shaped cartouche, the seal has bamboo leaves incised on its top.

•64
Irregular seal
China, Ch'ing dynasty (1644-1912), 7.9 cm. h.
This naturally shaped piece of bamboo is said to resemble Shou Lao, the god of longevity (see 4).

•65
Irregular seal
China, Ch'ing dynasty (1644-1912), 11.3 cm. h.
Full advantage was taken of a sculptural piece of twisted and scarred bamboo in forming this seal.

66
Covered box with flowers
China, Ch'ing dynasty (1644-1912), 2.9 x 6.4 cm.

67

68

72

•67
Brushpot with scenes of animals and plants
Korea, 18th or 19th century, wooden base attached, 14.2 cm. h.
The brushpot's seven lobes are carved with scenes of flora and fauna. Deer, cranes, ducks, fish, bamboo, pine, plum, and chrysanthemums are among the subjects represented.

•68
Brushpot with three compartments
Korea, 18th or 19th century, wooden base attached, 16.5 cm. h.
Three sections of culm are joined together to form this brushpot. Cranes, a bat, and various plants are depicted on its exterior.
Published: Lutz, "Bamboo Brushpots," p. 32.

69
Box with double compartment
Korea, 18th or 19th century, bamboo, wood, and metal, 7 x 16 x 11.5 cm.
Perhaps used to hold writing articles, this box is veneered with bamboo strips and fitted with metal hinges, corner reinforcements, feet, and a clasp.

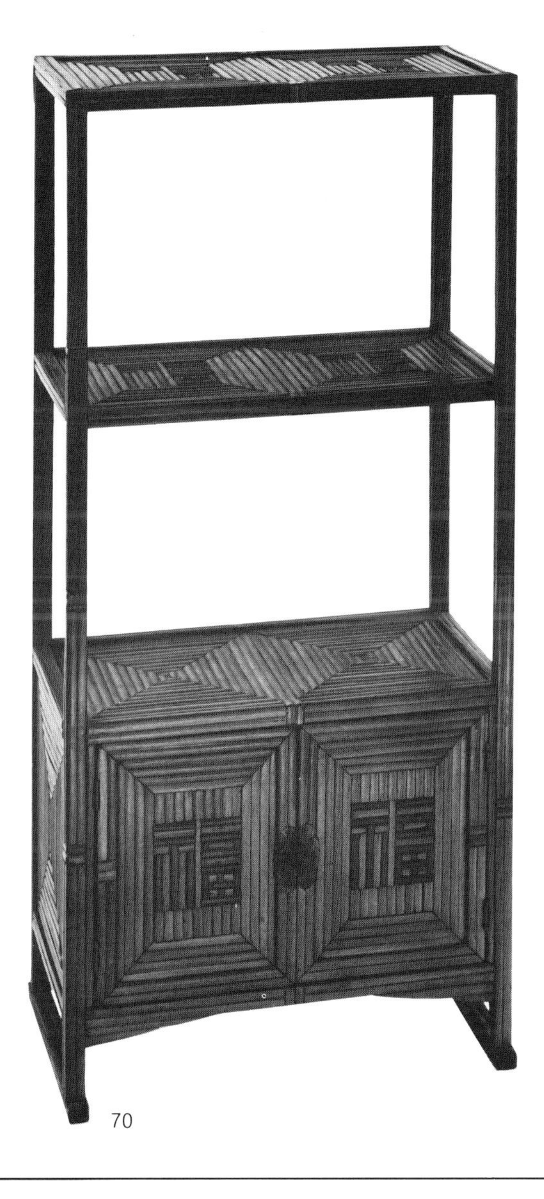
70

•70
Display shelf
Korea, 18th or 19th century, bamboo, wood, and metal, 104.7 x 48.5 x 28.3 cm.
This three-level shelf with storage compartment may have been used to furnish a scholar's studio. Darker strips of bamboo have been used to incorporate characters meaning "good fortune" into the veneer decoration.

•71
Brushpot with rooster and hen
signed: Sankōsai
Japan, Edo period (1615-1868), lacquer on bamboo, 15 cm. h.
A rooster, hen, and small birds with bamboo and pine are shown in slight relief.
Published: Lutz, "Bamboo Brushpots," cover.
color plate p. 12

•72
Wristrest with banana and chrysanthemum
signed: Tsuji Manpo
Japan, 20th century, 22.9 x 7.3 cm.

73
Wristrest with plum
Japan, 20th century, 25.3 x 6.6 cm.

74

76

•74
Inkstick rest with two alder leaves and catkin
Japan, Edo period (1615-1868), 1.4 x 4.5 cm.
75
Inkstick rest with leaf
Japan, Edo period (1615-1868), 1.5 x 7.4 cm.

•76
Inkstone set
signed: Kizan
Japan, 19th century, Tablet 9 x 6 cm., *netsuke* 6.9 cm. l., *ojime* .9 cm. l.
This portable inkstone set includes a tablet for grinding ink, an *ojime*, and *netsuke*. They are all made of bamboo.
Published: Austin and Ueda, *Bamboo*, p. 185.

•77
Writing set with dragon
signed: Haruki-sanjin
Japan, 1814, bamboo, gourd, and lacquer.
Brushholder 28.2 cm. l., ink case 7.6 cm. h., *ojime* 1.9 cm. dia.
Portable writing sets (*yatate*) such as this were used to carry a brush and ink. This particular example has an openwork bamboo brushholder and an ink case made from a gourd. They are joined by a cord with a red lacquer *ojime*.

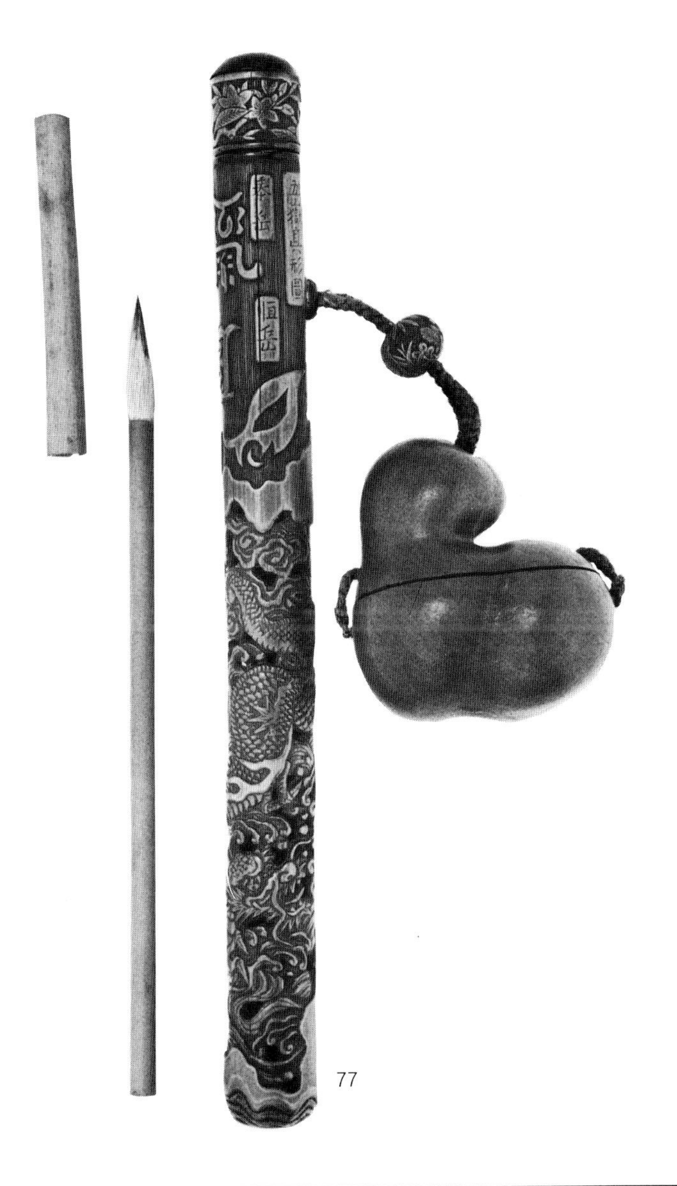

77

80

78
Writing set with cherry blossoms and duck
Japan, 19th century, lacquered bamboo, ivory, and gold.
Brush and holder 17.7 cm. l., ink case 4 cm. h.,
ojime .9 cm. dia.
The "sesame seed" bamboo brushholder is decorated with falling cherry blossoms; a pair of swimming ducks is depicted on the ink case. The set is fitted with an ivory brush and gold *ojime*.

79
Writing set
Japan, 19th century. Brush and holder 18.2 cm. l.,
ink case 2.5 cm. h., *ojime* .8 cm. l.

•80
Seal-ink box with pines
signed: Taisai (Koma school)
Japan, 19th century, lacquered bamboo, 3.4 x 8.5 x 8.5 cm.
This shallow, square box holds the reddish ink used to print seals.

81
Letter carrier
Japan, Edo period (1615-1868), 22 x 8.1 cm.

82
Letter carrier with crabs in a lotus pond
Japan, Edo period (1615-1868), bamboo and metal,
22.5 cm. l.

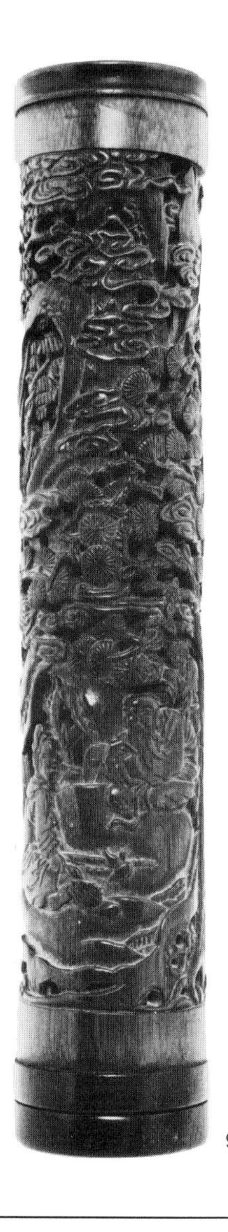

94

90

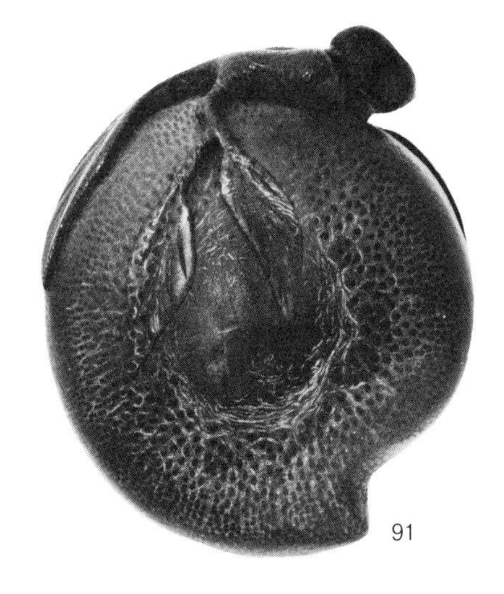

91

83
Letter carrier in the shape of a sword
Japan, Edo period (1615-1868), bamboo, wood, ivory, and mother-of-pearl, 32.9 cm. l.
Shaped as a sword and sheath, this letter carrier is carved, incised, and inlaid with butterfly motifs.

84
Letter opener
Japan, 19th century, 20.9 cm. 1.
The letter opener is enclosed in a sheath which is carved to resemble a *koto*.

85
Brush
Japan, 20th century, 38.5 cm. l.
This brush is made from a section of bamboo, one end of which was buried in the ground until it decomposed into its natural filaments.

Vessels and Containers

•86
Cup with pine tree and cranes
China, 18th century, 7.6 cm. h.
The shape of this cup is similar to cups carved from rhinoceros horn, which have an elliptical lip, everted mouth, and deep walls.
Published: Austin and Ueda, *Bamboo,* p. 175.

87
Cup with lotus flowers
China, 18th century, 4.2 cm. h.

86

88
Cup with grasshopper and flowers
China, 18th or 19th century, 8.8 cm. h.

89
Cup with archaic script
China, 19th century, 5.3 cm. h.

•90
Tea scoop with monkey and herons
China, 18th or 19th century, 14.1 x 6.9 cm.
The monkey in a peach tree and the herons and fish in a lotus pond are carved on the interior of a section of culm.

•91
Snuff bottle in the shape of a peach
China, 18th century, 5.9 cm. h.

92
Incense stick container with *lohan* and lady with attendants
China, 18th century, wooden base and cap attached, 15.3 cm. h.

93
Incense stick container with horsemen and pavilion
China, 18th century, ivory base and cap attached, 19.1 cm. h.

•94
Incense stick container with men seated in a landscape
China, 18th or 19th century, wooden base and cap attached, 18.2 cm. h.

96

•95
Vase with checkered weave
China, 14th century or later, lacquered bamboo and wood, 28.6 x 24.8 cm.
illustrated p. 18

•96
Dining set in gourd-shaped container
China, Ch'ing dynasty (1644-1912), lacquered bamboo and metal. Container 59 cm. h.
A dining set complete with cups, bowls, and plates is nested into a hinged, gourd-shaped carrying case.

97
Set of four plates
China, Ch'ing dynasty (1644-1912), lacquered bamboo, 13.5 cm. dia.
The dishes have a rim of red-lacquered bamboo in a checkered weave surrounding a scene painted in gold lacquer on black.

98
Rectangular box
China, Ch'ing dynasty (1644-1912), bamboo, wood, and metal, 13 x 37.1 x 14.6 cm.
Constructed of a wooden frame and bottom, this box with a hinged lid has panels of twilled bamboo.

99
Rectangular basket
China, Ch'ing dynasty (1644-1912), bamboo, wood, cane, and metal, 40 x 41.2 x 23.5 cm.
This single-compartment, covered container has a wooden base and a twill-weave bamboo foundation. It is trimmed with bamboo strips and metal fittings, and its handle is wrapped with cane.

100

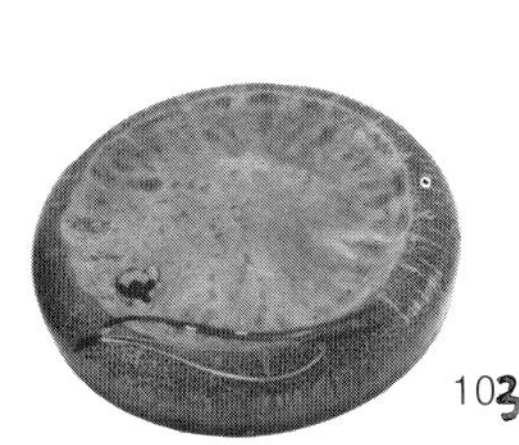

103

101

•100
Circular basket with two compartments
China, 1879, bamboo, wood, and metal, 51.4 x 41.2 cm.
The twilled design on the cover of this basket identifies its owner as a second-ranking wife.

•101
Tea bowl
Japan, 19th century, 6.4 cm. h.
The principal utensil of every tea ceremony is the tea bowl (*chawan*), which is ceramic as a rule; thus, this bamboo bowl has a carved spiral under its foot as if it had been trimmed by a potter.

•102
Tea caddy
Japan, 19th century, bamboo with lacquered interior, 5.1 cm. h.
A container for thin tea (*usu-cha*), this tea caddy is in the shape of a jujube fruit (*natsume*).
color plate p. 16

•103
Tea caddy with the Three Friends
signed: [Shibata] Zeshin (1807-1891)
Japan, 19th century, lacquered bamboo, 7 cm. h.
Made of bamboo and decorated with a plum branch and pine needles, this tea caddy represents the Three Friends, a symbol of happiness and good fortune.

104

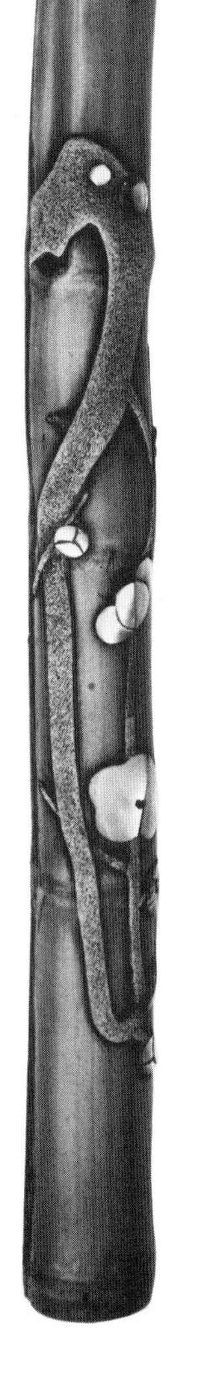

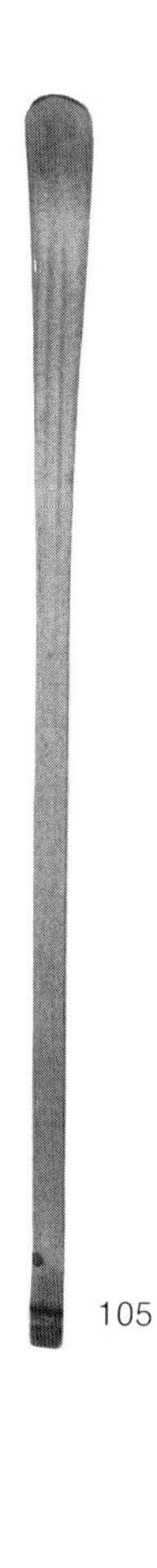

105

106

•104
Tea caddy
signed: Gengensai
Japan, early 20th century, bamboo with lacquered interior, 5.9 cm. h.
This tea caddy is woven from a square base to a circular shoulder and fitted with a round cap.

•105
Tea scoop and case
Japan, Edo period (1615-1868), bamboo, wood, lead, and mother-of-pearl. Scoop 19.2 cm. l., case 25.6 cm. l.
A tea scoop (*chashaku*) such as this is used to draw powdered tea from its container. This scoop's bamboo case (*tomo-zutsu*) is unusual in being decorated with a plum tree of inlaid lead and mother-of-pearl. On the case is a seal of Ogata Kōrin (1658-1716).

•106
Tea scoop in the shape of a cabbage leaf
Japan, 19th century, 14.2 cm. l.

107
Tea scoop with pomegranates
signed: Tōkoku
Japan, 19th century, 22.3 x 5.7 cm.

•108
Water ladle
Japan, 19th century, 28.1 cm. l.
Usually the water ladle (*hishaku*) is made of separate pieces of bamboo for the cup and handle. In this case, it has been fashioned from a culm and its natural branches.

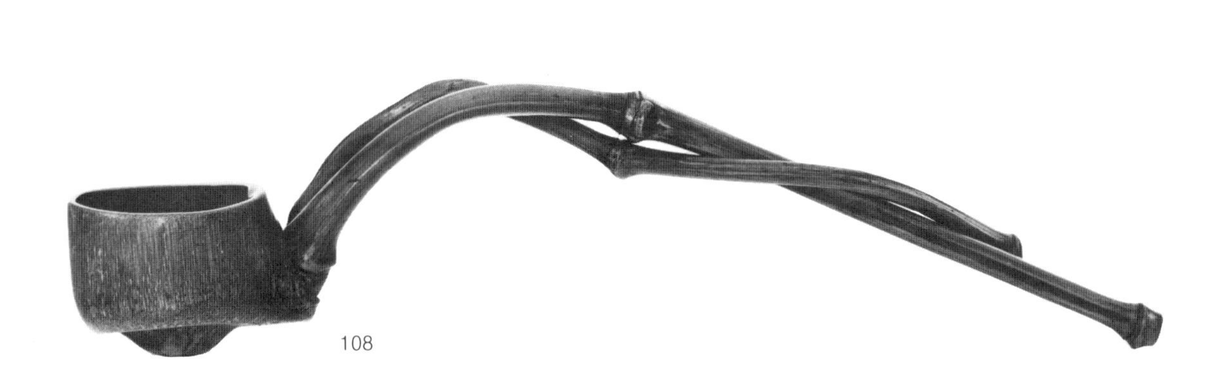
108

111

109
Tea whisk and case
Japan, 19th or 20th century. Whisk 11.5 cm. h., case 8.1 cm. h.
This tea whisk (*chasen*) of smoked bamboo is used to blend powdered tea and hot water.

110
Lid rest
Japan, 20th century, lacquered bamboo and metal, 5.5 cm. h.

•111
Water jar
Japan, 19th century, bamboo with lacquered interior, 15.2 cm. h.
Generally the water jar (*mizusashi*) that the host carries into the tea room is made of ceramic, lacquer, or wood. Made of bamboo, this water jar is quite light in weight.

112
Water bowl
Japan, 20th century, lacquered bamboo, 8.9 cm. h.

113
Square teapot
Japan, 19th century, 5.7 cm. h.

114
Round teapot
Japan, 19th century, 5 cm. h.

•115
Teapot in the shape of a lotus plant
Japan, late 19th century, 5.7 cm. h.
The simulated insect holes on the lid act as a steam vent for the teapot.

115

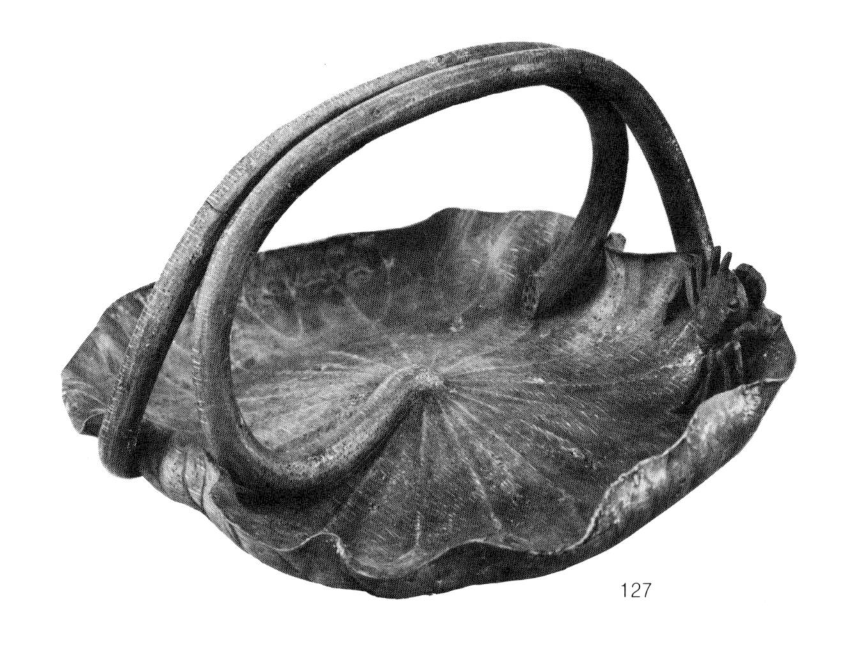
127

116
Teapot rest in the shape of a lotus leaf
Japan, 19th century, 12 cm. dia.

117
Teapot rest
Japan, 19th century, 12.5 cm. dia.
Made from a cross section of rhizome, this teapot rest is encircled by radiating roots. It has an old lacquer repair.

118
Circular coaster
signed: Shōkosai I
Japan, mid-19th century, 10 cm. dia.

119
Rectangular tray
signed: Gengensai
Japan, early 20th century, 2.6 x 20.6 x 17.3 cm.
The character for *fuku* ("good fortune") is woven into the bottom of the tray.

120
Rectangular box
signed: Chikuunsai
Japan, early 20th century, 6.6 x 14.2 x 10.2 cm.
The characters for *fukuju* ("prosperity and longevity") are interwoven on the box lid.

121
Box with Jittoku
Japan, Edo period (1615-1868), lacquered bamboo and wood, 5.7 cm. h.
The popular Zen figure Jittoku is identified by the broom lying in front of him. He performed menial tasks around temples and is usually shown with his companion, Kanzan (see 190).

128

123

122
Box with frog in a lotus pond
Japan, Edo period (1615-1868), lacquered bamboo, 5.4 x 10.8 cm.

•123
Box with radish
Japan, Edo period (1615-1868), lacquered bamboo and wood, 5.6 cm. h.

124
Box in the shape of a lotus plant
Japan, 19th century, 8.9 x 14.9 x 12 cm.

125
Sake ewer in the shape of a lotus plant
Japan, 19th century, lacquered bamboo, 9.2 x 22 x 16.8 cm.

126
Dish in the shape of a lotus plant
Japan, 19th century, 7.5 x 17 cm.

•127
Tray with crab on a lotus plant
Japan, 19th century, 11.7 x 21 cm.

•128
Tray with the Three Friends
signed: Chikusen
Japan, 19th century, lacquered bamboo and wood, 3.6 x 32.2 cm.
Five curved sections of culm form the walls of this plum-blossom-shaped tray. Its interior is lacquered red and decorated with a gold pine needle design.

129

130

132

•129
Napkin holder with two men on a basket
Japan, 19th century, 6 cm. h.

•130
Napkin holder with two mice on a winnower
Japan, 19th century, 4.3 cm. h.

131
Napkin holder in the shape of a Chinese flower basket
signed: Gengensai
Japan, early 20th century, 5.1 cm. h.

•132
Basket for outdoor tea ceremony
Japan, 19th century, bamboo, cane, and cloth, 24.8 cm. h.

133
Basket for outdoor tea ceremony
signed: Shōkosai III
Japan, early 20th century, bamboo, cane, metal, and cloth, 18.7 x 21 x 12.7 cm.
Rising from a hexagonal base to an oval rim with hinged lid, this basket includes an interior tray.

•134
Cabinet for tea ceremony utensils
Japan, 19th century, bamboo, wood, metal, stone, and mother-of-pearl, 52 x 40 x 25.7 cm.
Inlaid lacquer panels from China were adapted as the two doors for this cabinet. Inside there are two shelves.
color plate p. 12

135

138

•135
Cabinet for tea ceremony utensils
signed: Shōkosai IV
Japan, mid-20th century, 42.8 x 31.5 x 27 cm.

136
Basket for charcoal
signed: Waichisai
Japan, 20th century, bamboo and lacquered paper, 12 x 26.5 x 26.5 cm.

137
Basket for smoking accessories
signed: Shōkosai III
Japan, early 20th century, 17.6 x 33.6 x 22.8 cm.

•138
Smoking set
Japan, 19th century, bamboo and metal. Pipe 19.9 cm. l., pipe case 21.2 cm. l, tobacco case 4.5 cm. l., *ojime* 1.5 cm. l.

139
Pipe case with Daruma
Japan, late 19th or early 20th century, bamboo, wood, and metal, 22.4 cm. l.
Daruma (Sanskrit, *Bodhidharma*) is a popular figure in Japan. A Buddhist patriarch, he is the legendary founder of the Zen sect.

140
Pipe case with the White-Robed Kannon
Japan, late 19th or early 20th century, bamboo and ivory, 21 cm. l.

142

147

150

141
Pipe case
Japan, late 19th or early 20th century, bamboo and ivory, 22.8 cm. l.

•142
Incense container with cricket
Japan, 19th century, lacquered bamboo and mother-of-pearl, 2.1 x 7.1 cm.
Incense containers (*kogō*) were used in the tea ceremony, either displayed in the alcove or handed to guests for inspection.

143
Incense container
Japan, 19th century, 2.7 x 7.1 cm.

144
Incense dish with overlapping leaves
Japan, 19th century, 2.2 x 9.2 cm.
This incense dish is carved from the diaphragm of a culm.

145
Incense stick container with grapevine and wasp
signed: Ikkokusai
Japan, 19th century, lacquered bamboo, 37.9 cm. l.

146
Incense stick container with flowers
signed: Tōkoku
Japan, late 19th or early 20th century, bamboo and ivory, 34.7 cm. l.
This tube for holding incense sticks is carved from a type of bamboo which is naturally squarish in section.

151

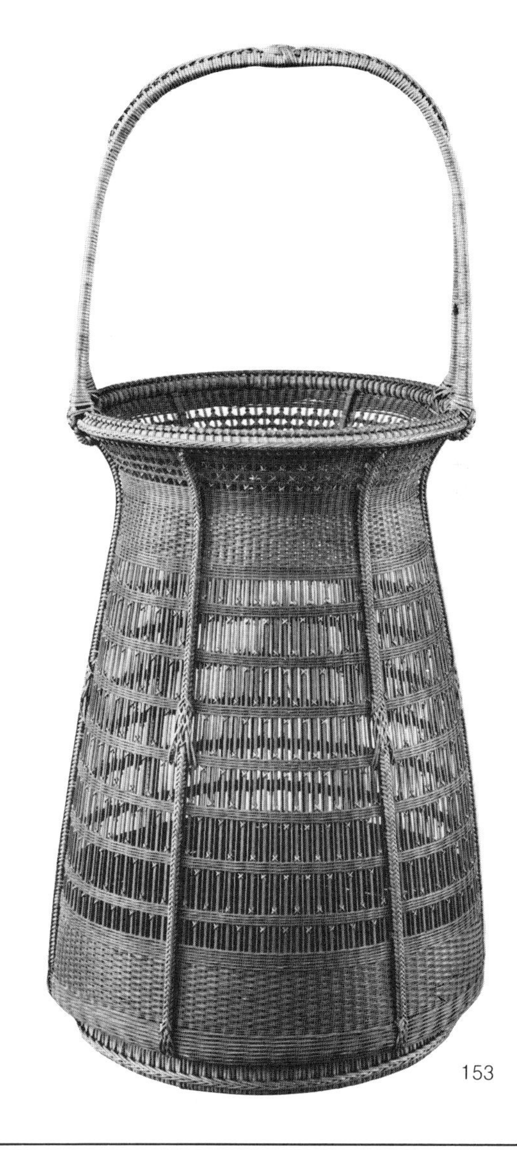

153

•147
Incense stick container with *gigaku* mask
signed: Tetsurō
Japan, early 20th century, 26.7 cm. l.
Gigaku is the oldest tradition of dance imported to Japan from the Asian continent. It was practiced at temples in Nara during the 7th and 8th centuries. The mask depicted on the incense container is from the temple collection of Hōryūji.

148
Incense stick container with bamboo leaf and calligraphy
Japan, early 20th century, lacquered bamboo and ivory, 58.5 cm. l.
A rare type of bamboo from northern Kyūshū was used for this incense container. The space between nodes of the culm is particularly long.

149
Circular basket with handle
signed: Shōkosai I
Japan, mid-19th century, 15.2 cm. h.

•150
Basket with two ear handles
signed: Shōkosai I
Japan, mid-19th century, 13.7 cm. h.

•151
Flower basket
signed: Hōryūsai
Japan, late 19th century, 59.3 cm. h.
This basket is woven with a double layer of bamboo — a fine mesh on the outside and a coarser weave for the interior.

155

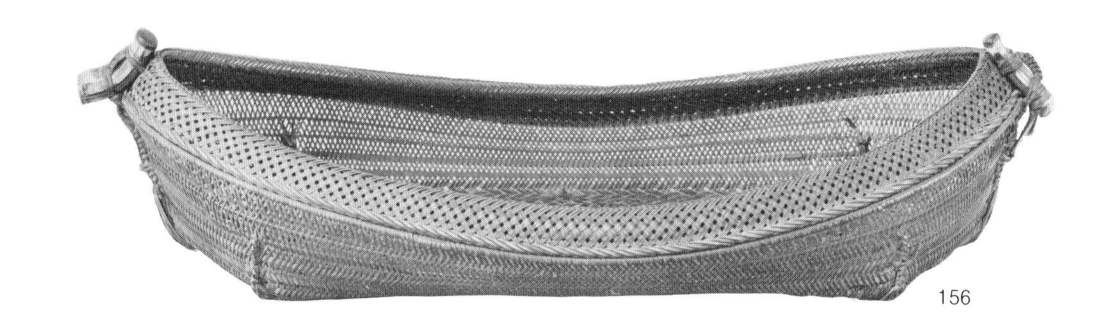

156

152
Flower basket
signed: Chikuunsai II
Japan, 1947, 40.6 cm. h.
The walls of this tall, squarish basket are done in a hexagonal weave. A bamboo insert acts as a water receptacle.

•153
Flower basket
Japan, 20th century, 41 cm. h.
This finely woven basket is laced in a technique similar to one used for Japanese armor.

•154
Flower basket
Japan, early 20th century, 14.7 cm. h.
Ikebana arrangement: Mona Lutz (Ohara school)
color plate p. 13

•155
Hanging flower basket
Japan, 20th century, 25.4 cm. h.

•156
Boat-shaped flower basket
signed: Banshōsai
Japan, 19th century, bamboo with metal insert, 12.4 x 53.3 x 27.3 cm.

•157
Flower container
Japan, Edo period (1615-1868), 21.6 cm. h.
This flower container is based on a model created by Sen no Rikyū (1521-1591) during the spring of 1590. Rikyū's bamboo container, which split while drying, was named *Onjōji* after the famous cracked bronze bell at that temple.

158
Flower container
Japan, Edo period (1615-1868), 22.6 cm. h.

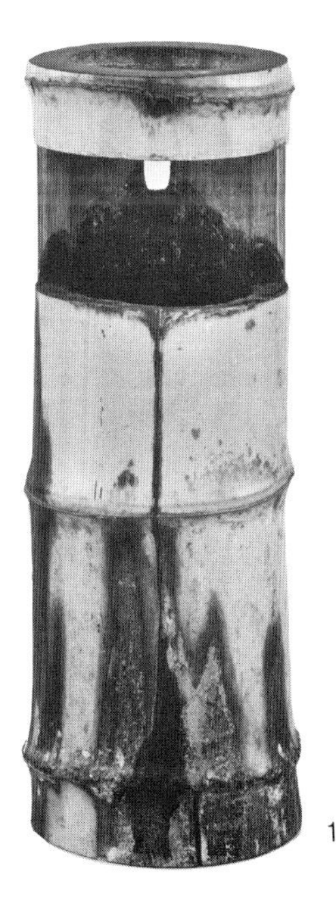
157

159

164

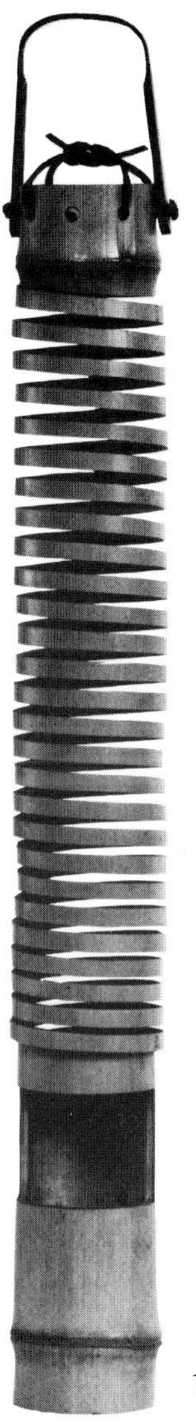
165

•159
Flower container with crab and lotus plant
signed: Ippō
Japan, Edo period (1615-1868), 16.1 cm. h.

160
Flower container with calligraphy
signed: [Sakai] Hōitsu (1761-1828)
Japan, 19th century, lacquered bamboo, 24.4 cm. h.
The elegant, cursive calligraphy is written in gold lacquer on the back of the container.

161
Flower container with calligraphy
signed: Issō
Japan, 19th century, 23 cm. h.

•162
Hanging flower container
Japan, Edo period (1615-1868), 29.8 cm. h.
Designed to be suspended, this flower container is made from a rhizome. *illustrated p. 10*

163
Faceted flower container
signed: Shōfūan Teishō
Japan, 20th century, 17.4 cm. h.

•164
Flower container
signed: Shōkosai II
Japan, mid-20th century, 49 cm. h.

•165
Hanging flower container "Mountain Stream" *(Keiryū)*
signed: Shihōkan Sōhen
Japan, 18th or 19th century, 55 cm. h. unextended.
The culm has been cut so that it falls into a spiraling coil when the flower container is suspended.

166

167

170

171

Utility and Adornment

•166
Earspoon container with hunting scene
China, 18th century, bamboo and ivory, 11.9 cm. h.

•167
Case for spectacles with gentlemen and attendant
China, 1767 or 1827, 16.2 cm. h.
The carving has been done on the interior skin of the culm, which has been laminated to the exterior of the case.

•168
Hatstand with calligraphy and flowers
China, 19th century, bamboo and wood, 30.5 cm. h.

169
Fan with figures in a landscape
China, 19th century, bamboo and wood, 41 cm. h.

•170
Folding fan with calligraphy on frame
China, late 18th or early 19th century, bamboo and paper, 34.3 cm. h.

•171
Folding fan with calligraphy on frame
China, 19th century, bamboo and paper, 33.9 cm. h.

172
Carrying case for a cricket
China, 19th century, 15.4 cm. l.

173
Seat pad
China, 20th century, bamboo and wood, 34.7 x 22.4 cm.

168

177

178

179

174
Five-rung ladder
China, 20th century, 165.3 cm. h.

175
Quiver with bamboo and grapevines
Korea, 18th or 19th century, 97.9 cm. l.

176
Lantern
Korea, 18th or 19th century, bamboo, wood, metal, and paper, 24.7 cm. h. without handle.

•177
Compass and carrying case
Korea, 20th century, bamboo, metal, and paper, 3.7 cm. dia.
With proper tension on the cord, the case opens to reveal a small drawer containing a compass.

•178
Inrō* in the shape of a lantern, *ojime,* and *netsuke
Japan, 18th century, bamboo and wood.
Inrō 9.3 cm. h., *ojime* 1.7 cm. l., *netsuke* 3.5 cm. h.
The top of the lantern acts as a cap for the *inrō,* whose cruciform cross section may be a covert Christian symbol.

•179
***Inrō* with figures in a landscape**
Japan, 18th century, 7.2 cm. h.
An oval seal fits into the top of the *inrō,* and the lower of its two compartments holds the seal ink.

180
Inrō with figures in a landscape
Japan, 18th century, bamboo and metal, 3.9 cm. h.

181

182

187

•181
Inrō with Daruma and lion
signed: Bunyū
Japan, 18th or 19th century, 9.1 cm. h.

•182
Inrō with checkered inlay
Japan, 19th century, 7.7 cm. h.
One side of the *inrō* slips open to reveal five small drawers.
Published: Austin and Ueda, *Bamboo,* p. 187.

183
Inrō with flying bird
Japan, Edo period (1615-1868), lacquered bamboo and metal, 7.6 cm. h.

184
Inrō with hawk and bird in a landscape
Japan, Edo period (1615-1868), lacquered bamboo and metal, 6.2 cm. h.

•185
Inrō with hatching chick, *ojime*
signed: [Shibata] Zeshin (1807-1891)
Japan, 19th century, lacquered bamboo and amber.
Inrō, 5.8 cm. h., *ojime* 1 cm. dia.
color plate p. 16

186
Inrō and *ojime*
Japan, Edo period (1615-1868), bamboo and metal.
Inrō 8.1 cm. l., *ojime* 1.4 cm. l.

190

195

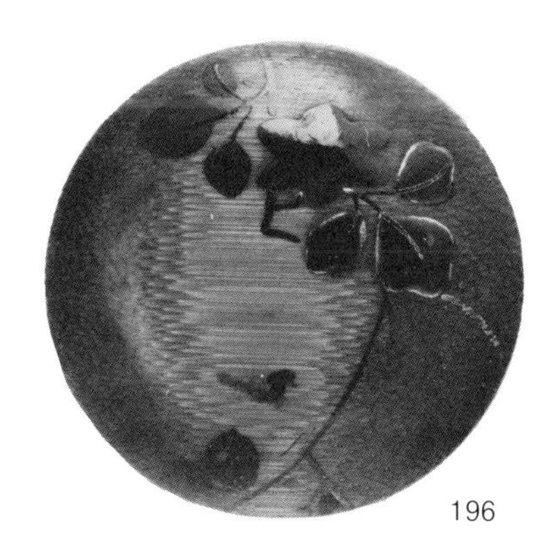
196

197

•187
Netsuke with owl, *ojime*
Japan, 19th century. *Netsuke* 5.1 cm. h., *ojime* 2.1 cm. l.

188
Netsuke with face
signed: Suihō-sanjin
Japan, 19th century, bamboo and wood, 9.1 cm. h.

189
Netsuke with two faces
Japan, Edo period (1615-1868), 3.5 cm. l.
Each end of the *netsuke* is carved with an open-mouthed face.

•190
Netsuke with Kanzan and Jittoku
Japan, Edo period (1615-1868), 5.4 cm. h.
The eccentric poet Kanzan (Chinese, *Han-shan)* and his friend Jittoku (Chinese, *Shih-te)* are shown with their respective attributes, a scroll and a broom.

191
Netsuke with pair of wrestlers
Japan, Edo period (1615-1868), 4.5 cm. h.

192
Netsuke with two animals
Japan, Edo period (1615-1868), 3.4 cm. h.
The two creatures, shown face-to-face, combine to form a bell.

193
Netsuke with pomegranates, *ojime*
Japan, Edo period (1615-1868). *Netsuke* 5 cm. h., *ojime* 1.5 cm. l.

194
Netsuke in the shape of a cup
Japan, Edo period (1615-1868), 1.9 cm. h.

200

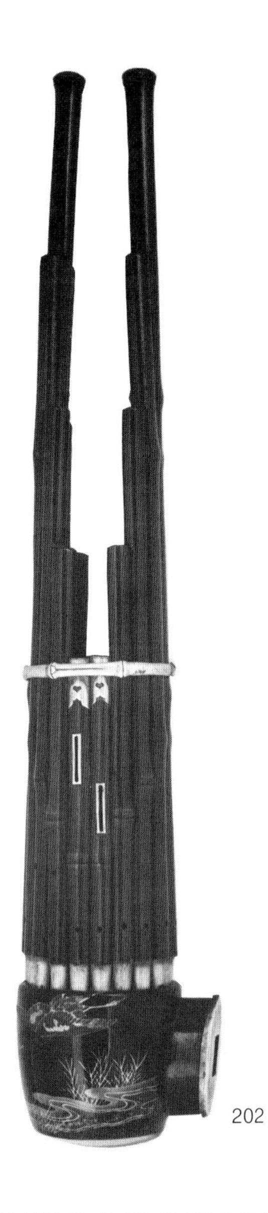

202

203

•195
Manjū netsuke with dragon
Japan, Edo period (1615-1868), 1.7 x 4.3 cm.

•196
Manjū netsuke with clematis
signed: Issai
Japan, Edo period (1615-1868), lacquered bamboo, 1.5 x 3.5 cm.

•197
Ojime with fly
Japan, Edo period (1615-1868), lacquered bamboo, 1.2 cm. h.

198
Sash clip with Sanskrit characters
Japan, 19th or 20th century, 1.2 x 4 x 3.7 cm.

199
Plaque with crab catching a toad
Japan, 18th or 19th century, 5.4 cm. h.

•200
Case for spectacles with Buddha, Shōki, and demon
Japan, 19th century, bamboo and metal, 18.2 cm. h.

201
Folding fan with plum trees on frame
Japan, 20th century, bamboo and paper, 33.2 cm. h.

•202
Mouth organ
Japan, 19th century, bamboo, lacquer, and metal, 45.4 cm. h.
This musical instrument *(shō),* made of various lengths of bamboo pipes bound together, is a kind of mouth organ.

205

207

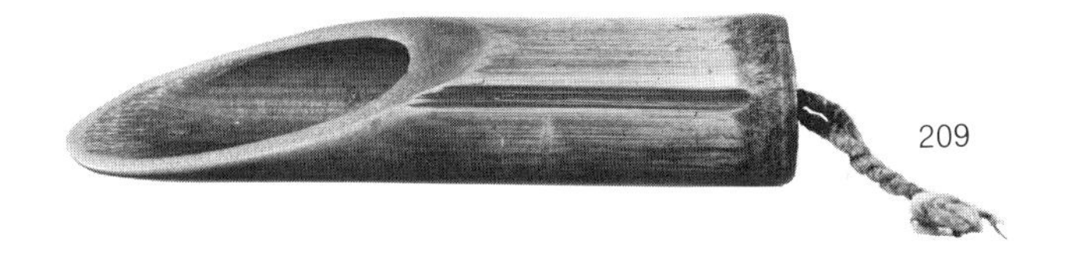

209

•203
Vertical flute and case
Japan, 18th or 19th century, lacquered bamboo.
Flute 18.2 cm. l., case 20.7 cm. l.

204
Transverse flute
Japan, 20th century, lacquered bamboo, 39.5 cm. l.

•205
Clapper
Japan, 18th or 19th century, 34.4 cm. l.

•206
Monk's hat
Japan, 19th or 20th century, 18.4 x 55.2 cm.

•207
Insect cage
Japan, 19th century, lacquered bamboo,
12 x 11.3 x 7.3 cm.

208
Compass and case
Japan, 19th century, bamboo, metal, and glass,
3.8 x 8 cm.

•209
Rice tester
Japan, 19th century, 14.7 cm. l.
By inserting the tester into flexibly woven sacks filled with rice, samples could be removed for examination.
Published: Austin and Ueda, *Bamboo,* p. 94.

210
Kettle hook
Japan, 19th century, bamboo and wood, 122.3 cm. l.
Used to suspend a kettle over the hearth, this *jizai* consists of a hanger, rod, and adjuster.

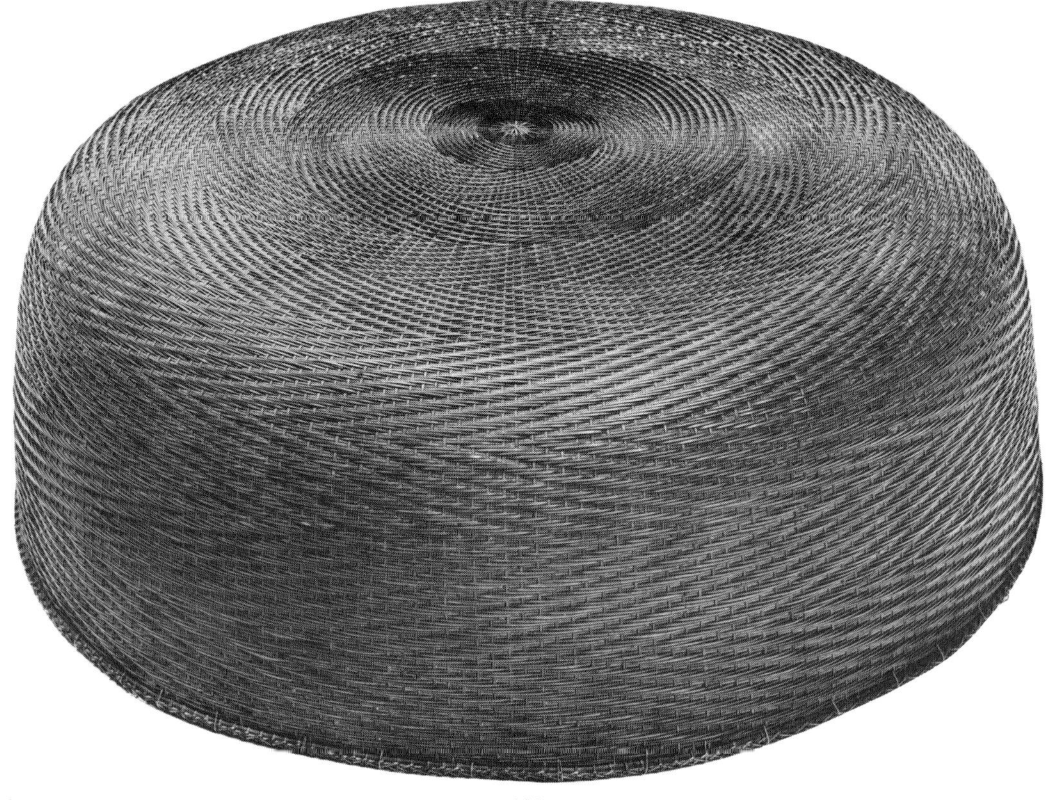

206

211
Basket for tea leaves
Japan, 19th century, 11.2 x 57 x 56.5 cm.

212
"Bamboo princess" or "Dutch wife"
Japan, 19th century, 109 cm. l.
An openwork bolster of bamboo such as this was used in a bed on hot nights so that air could circulate around the body.

213
Cane with flask
Japan, 19th century, bamboo, ivory, and metal, 96 cm. l.

214
Cane
Japan, 19th century, bamboo and metal, 104 cm. l.

215
Shade
Japan, 20th century, bamboo, metal, and cloth, 170.5 x 94.5 cm.

216
Shade with birds, flowers, and rock
Japan, Edo period (1615-1868), bamboo, silk, metal, and cloth, 99.5 x 136.5 cm.
Split sections of bamboo have been individually wrapped with colored silk so that when they are woven together, a pictorial design is created.

217
Latticework screen
Japan, 19th century, bamboo, wood, and leather, 45.8 x 120.5 cm. extended.
Made of black bamboo, the six panels are hinged together by thin leather straps.

220

218
Display board
Japan, 19th century, bamboo and wood,
1.9 x 48.2 x 29.8 cm.

219
Semicircular display stand
Japan, late 18th or early 19th century, bamboo
and lacquered wood, 7.2 x 28.6 x 11.7 cm.

•220
Display stand
Japan, 19th century, bamboo and lacquered wood,
42 x 45.4 x 27.2 cm.

221
Display stand
Japan, 19th century, bamboo and wood,
48.5 x 33.3 x 26.4 cm.

222
Oval stand for an incense burner
Japan, 19th century, 8.8 x 43.8 x 26.5 cm.

•223
Low table
Japan, 19th century, bamboo and wood,
14.5 x 45.1 x 30.3 cm.

224
Low table
Japan, 19th century, 11.7 x 73.5 x 29.5 cm.

225
Table
Japan, 19th century, bamboo, wood, and burl,
30 x 38.7 x 25.1 cm.

223

226
Table with cracked-ice fretwork
Japan, 19th century, bamboo and lacquered wood, 28.3 x 48.9 x 28 cm.

227
Flower arrangement set
Japan, 19th century, lacquered bamboo and metal. Case 19.5 x 5.8 cm.
The kit includes scissors and a knife enclosed within a carrying case.

228
Bamboo
Asai Tonan (1706-1782)
Japan, 18th century, ink on silk, 106.4 x 32 cm.
Asai Tonan was born in Owari province and lived in Kyoto. He was a doctor by profession but studied painting as a hobby. He was known for his use of bamboo as a subject. In both China and Japan, painting and calligraphy were closely related. Mastery of these arts was based largely upon the brushwork used in painting bamboo.

•229
Pipe
Thailand, 18th or 19th century, bamboo and metal, 27 cm. l.
illustrated p. 8

230
Cane
Thailand, 19th century, bamboo and metal, 93.2 cm. l.